# The Secret Lives of Numbers

## Numerals and Their Peculiarities in Mathematics and Beyond

Alfred S. Posamentier

Prometheus Books

Guilford, Connecticut

 **Prometheus Books**

An imprint of Globe Pequot, the trade division of The Rowman & Littlefield Publishing Group, Inc.
4501 Forbes Blvd., Ste. 200
Lanham, MD 20706
www.rowman.com

Distributed by NATIONAL BOOK NETWORK

British Library Cataloguing in Publication Information Available

**Library of Congress Cataloging-in-Publication Data**
Name: Posamentier, Alfred S., author.
Title: The secret lives of numbers : numerals and their peculiarities in mathematics and beyond / Alfred S. Posamentier.
Description: Lanham, MD : Rowman & Littlefield, [2022] | Summary: "The Secret Lives of Numbers takes readers on a journey through integers, considering their numerological assignments as well as their significance beyond mathematics and in the realm of popular culture"—Provided by publisher.
Identifiers: LCCN 2021053575 (print) | LCCN 2021053576 (ebook) | ISBN 9781633887602 (paperback) | ISBN 9781633887619 (epub)
Subjects: LCSH: Number concept. | Numerals. | Arithmetic—Foundations.
Classification: LCC QA141.15 .P675 2022 (print) | LCC QA141.15 (ebook) | DDC 513—dc23/eng/20220104
LC record available at https://lccn.loc.gov/2021053575
LC ebook record available at https://lccn.loc.gov/2021053576

♾️™ The paper used in this publication meets the minimum requirements of American National Standard for Information Sciences—Permanence of Paper for Printed Library Materials, ANSI/NISO Z39.48-1992

To Barbara for her support, patients, and inspiration.

To my children and grandchildren, whose future is unbounded: Lisa, David, Daniel, Max, Samuel, Jack, and Charles.

And in memory of my beloved parents, Alice and Ernest, who inspired me from day one!

# Contents

# Introduction

Long before today's technological world, we have been surrounded by numbers. We see numbers on automobile license plates, telephone numbers, addresses, weather reports as well as many commercial analyses. Yet we look at these numbers for their role as descriptors and not as an entity unto themselves. It is high time that we give the numerals and numbers special attention as they very often carry a great deal of hidden wonders. This book will take you on a journey through the numbers, considering their numerological assignments as well as their mathematical significance. For example, it is well-known in many circles that the number 13 carries a certain value of unluckiness with it, the phobia of which is often referred to as *triskaidekaphobia*. Many famous people have suffered from this phobia such as Napoleon, Mark Twain, and Franklin Delano Roosevelt, who was known to not invite guests to a party when there were 13 people in attendance. There are high-rise buildings that skip the 13th floor, just to mention a few aspects of this unlucky number. There are many explanations as to how the number 13 received this negative honor, which we will discuss in this book. From a mathematical point of view, the number 13 is the smallest prime number that when its digits are reversed is also a prime number. It is also honored with a place among the Fibonacci numbers and integral Pythagorean triples, as well as many other occurrences. This is just one example of a popular number that will be discussed (naturally, in greater detail) in our journey through the many interesting numbers that we will encounter.

The book addresses the numbers in order of size, beginning with single-digit numbers, followed by two-digit numbers, then three-digit numbers and then lots of strange numbers beyond the three-digit numbers. The book is intended for the general readership and will require no mathematics beyond the first two years of high school mathematics. In most cases, the mathematical significance of the numbers will be explained in arithmetic fashion.

All the units are be independent of one another which would allow the reader to search for favorite numbers and be able to discover the hidden significances both numerologically and mathematically. The book concludes with an "Special Number Characteristics" section that sheds light on general peculiarities of number relationships, some of which have been discussed throughout the book and others will have been pieced together from previous discussions. Above all, the aim of the book is to show the beauty and power that is so well hidden in our numbers with the hope that the reader will be motivated to embark on further investigations.

*1*

# Significant Numbers and What They Mean, Represent, and Exhibit

## THE NUMBER ZERO

Had we been writing 2,000 years ago about the number, which today we call *zero*, we would be at a loss. The concept in calculation, where an empty space was first used, can be found in the second century before the common era (BCE) with the Babylonians. However, the numeral 0 seems to have originated in India in the third century of the common era (CE) in the Bakhshali manuscript and was written in Sanskrit on birchbark leaves. It was called *sunya*, which meant blank or empty. When the numerical symbols that we use today traveled to the Arab world the zero was referred to as *sifir*, which then eventually morphed into *cipher* and *zero*.

The zero's big entrance into the Western world came largely from a book, *Liber Abaci*, published in 1202 by Leonardo of Pisa (c. 1170–c. 1240), who is better known as Fibonacci, where in the first sentence of the book he states: "The nine Indian figures are: 9, 8, 7, 6, 5, 4, 3, 2, 1. With these nine figures, and the sign zero which the Arabs called zephyr any number whatsoever is written, as is demonstrated." Fibonacci spent several years at the Pisan trading colony in Bugia, Algeria, where his father was the chief magistrate. This gave Fibonacci the opportunity to work with Arab mathematicians, where he first encountered these numerals. It is believed that this was the first appearance of these now-referred-to-as Hindu-Arabic numerals in Western Europe. This is where the zero made its great debut in our number system.

The zero plays a very important role in mathematics, not merely as a placeholder such as in the number 503, where it tells us there are no tens in this number, but also in arithmetic. For example, 0 enables us to represent the number 1 in a variety of ways: $3^0 = 1$, $18^0 = 1$, $5{,}280^0 = 1$. In other words, any real number to the zero power is equal to 1—including $0^0 = 1$. More about this later.

1

Multiplying any number by zero results in a zero. However, division by zero encounters a significant problem. When we check a division problem, we multiply the quotient by the divisor. If the result of this multiplication is the dividend, then we know that the division was done correctly. For example, if the divisor is zero as in this situation: $\dfrac{18}{0} = x$, a problem arises. In checking this division, we would multiply the divisor 0 by the quotient $x$ and we would never be able get 18. Therefore, mathematicians have left division by zero as undefined, or to put it another way, inadmissible.

Just to show what can happen if division by zero were permissible, we would be able to prove some rather ridiculous results. For example, we would be able to prove that $1 = 0$. For example, consider the following: we begin with our given information, namely, that $x = 0$. We then multiply both sides of this equation by $x - 1$ to get $x(x - 1) = 0(x - 1) = 0$. Now dividing both sides by $x$, leaves us with $x - 1 = 0$, which, in turn, tells us that $x = 1$. However, remember we began with $x = 0$. Therefore, we would then conclude that $1 = 0$ because both are equal to $x$. Absurd! Where did we go wrong? We divided by zero when we divided both sides of the equation by $x$, which we knew from the outset was equal to 0. Division by zero is, therefore, not permitted in mathematics, as it will lead us to silly conclusions, such as the one we just experienced.

Zero also plays a significant role in the sciences and culture. For example, although we have the year 1 BCE and the year 1 CE, there is no zero year. The zero also plays an important role in telephoning, in some countries to telephone abroad, you begin with one or two zeros; this varies according to country. On the Celsius scale, zero is the freezing point of water. In physics the lowest form of energy is denoted as the zero—point energy.

In mathematics, zero has upset ancient cultures because, as we noted earlier, multiplying a number by zero always resulted in zero, and when added or subtracted from a number left the number unchanged. This leaves us with a mathematical conundrum. We know that $x^0 = 1$, when $x \neq 0$, and that $0^x = 0$ when $x \neq 0$, then what is the value of $0^0$? In most aspects of mathematics, it is left as undefined, as we mentioned earlier. However, in some fields such as algebra and combinatorics the agreed–upon value is that $0^0 = 1$. Just to further expose the strange handling of the number zero, consider 0! (which is read as "zero factorial") and is also specially defined. We know that $n$-factorial is defined as $n! = 1 \times 2 \times 3 \times 4 \times \ldots \times n$. However, when $n = 0$, we define $0! = 1$. Although the zero entered into our mathematics history relatively late, it has proven to be a topic so rich that several books have been written about zero.

## THE NUMBER 1

We begin with the English word for the number 1, *one*, which is clearly not pronounced the way it is spelled. It is believed to have evolved from the English word *an*, which probably took its definition from the ancient Germanic word *ainaz*, which in today's German has evolved to the word *eins*, meaning one.

The number 1 is, of course, the first of our counting numbers, known as the *natural numbers*. It is also the first square number, the first triangular number, the first cubic number as well as the first Fibonacci number, and so it goes taking its initial place in various number collections.

When we initially encounter the number 1, we ask ourselves is this an even or an odd number? And is this a prime number or a composite number? By definition, an even number is one that can be split into two equal integers, which is not possible with the number 1; therefore, we would categorize it as an odd number. A prime number is usually defined as a number that has exactly two factors, the number 1 and another integer. This is not the case for the number 1, which only has one factor, namely, 1. Therefore, the number 1 is not a prime number. The number 1 does play a significant role in probability, where when the probability of an event is certain, it has a probability of 1.

Naturally, the number 1 has played a significant role in cultural history, where ancient civilizations have depicted values as the collection of ones. However, the Pythagoreans rejected the number 1 because multiplying any number by 1 did not change the original number. Also dividing any number by 1 leaves the original number unchanged. Yet, today the number 1 is often used as a ranking, 1 CE refers to the first year of the common era, the first month of the year, +1 is the code for international telephone calls to the North American continent, and so it goes that the initial number of our natural numbers takes on a multitude of important roles.

The Irrepressible Number 1 is one way to consider the following technique. There are times when we refer to beauty in nature as magical. Is magic beautiful? Some feel that when something is truly surprising and "neat" it is beautiful. From that standpoint, we will show a seemingly "magical" property in mathematics. This is one that has baffled mathematicians for many years and still no one knows why it happens. Try it, you'll like it.

We begin by asking you to follow two rules as you work with any *arbitrarily* selected number.

*If a given number is odd, then multiply it by 3 and add 1.*
*If the number is even, then simply divide the number by 2.*
*Regardless of the number originally selected, the result will always be 1.*

Let's try it for the *arbitrarily selected* number **12**.

Because 12 is even, therefore, divide by 2 to get 6.

Because 6 is also even, so we again divide by 2 to get 3.

Because 3 is odd, therefore, multiply by 3 and add 1 to get 10.

The number 10 is even, so we simply divide by 2 to get 5.

Because 5 is odd, so we multiply by 3 and add 1 to get 16.

Because 16 is even, so we divide by 2 to get 8.

Because 8 is even, so we divide by 2 to get 4.

Because 4 is even, so we divide by 2 to get 2.

Because 2 is even, so we divide by 2 to get 1.

No matter which number we begin with (here we started with 12) we will eventually get to 1. This is truly remarkable! Try it for some other numbers to convince yourself that it really does work. Had we started with 17 as our arbitrarily selected number we would have required 13 steps to reach the number 1. Starting with the number 43 will require 27 steps.

Does this really work for all numbers? This is a question that has concerned mathematicians since the 1930s, and to date no answer has been found, despite monetary rewards having been offered for a proof of this conjecture. Most recently (using computers) this problem, known in the literature as the "$3n + 1$ Problem" has been shown to be true for the numbers up to $10^{18} - 1$. The longest progression of steps so far established is for the number 931,386,509,544,713,451, which requires 2,283 steps to reach the number 1.

For those who have been turned on by this curious number property, we offer you a schematic (Figure 1) that shows the sequence of start numbers 1–20. The bold numbers (1–20) can be starting points for your progression following the preceding rules.

Notice that you will always end up with the final loop of 4-2-1. That is, when you reach 4 you will always get to the 1 and then were you to try to continue after having arrived at the 1, you will always get back to the 1 because by applying the rule you continue in the loop: 4-2-1.

We don't want to discourage inspection of this curiosity, but we want to warn you not to get frustrated if you cannot prove that it is true in all cases, for the best mathematical minds have not been able to do this for the better part of a century!

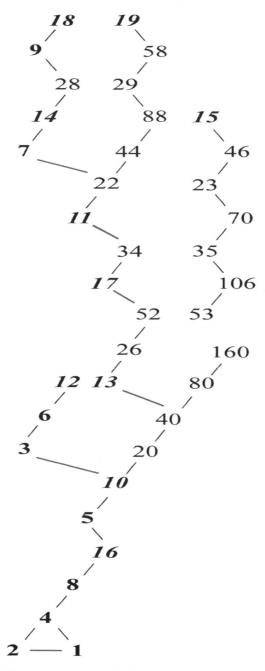

**Figure 1 Sequence of initial numbers 1–20 of "3n + 1."**

## THE NUMBER 1.618... (THE GOLDEN RATIO)

Few mathematical concepts have as great an impact on as many aspects of our visual and intellectual lives as the Golden Ratio, which is approximately equal to **1.618**03398874989484820458683436566. We begin by defining this famous ratio. The Golden Ratio evolves from of the lengths of the two parts of a line segment that forms the following proportion, where the longer segment (*L*) is to the shorter segment (*S*) as the entire original segment (*L* + *S*) is to the longer segment (*L*). Symbolically, this is written as $\dfrac{L}{S} = \dfrac{L+S}{L}$, which we see in Figure 2.

L                                              S

A                                P                                B

**Figure 2**

This is called the *Golden Ratio* or the *Golden Section*, in the latter case we are referring to the "sectioning" or partitioning of a line segment. Especially in geometry, where a *Golden Rectangle* is constructed from this ratio as the side length relationship. The Golden Rectangle is so named because of its apparent beauty, which has been established through many psychological experiments with human subjects expressing their perception of its beauty. The name of this ratio has evolved over centuries largely because it appears in architecture. For example, the Parthenon in Athens, Greece exhibits the Golden Ratio as shown in Figure 3.

**Figure 3**

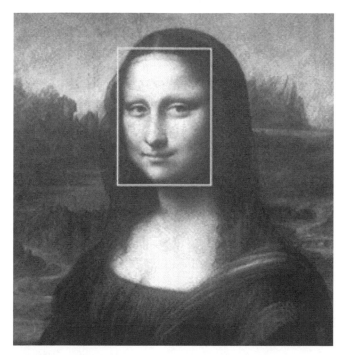

**Figure 4**

In art, the Golden Ratio used sometimes intentionally and sometimes artistically. For example, Leonardo da Vinci's Mona Lisa's face can be captured in a Golden Rectangle as shown in Figure 4. The Golden Ratio also appears in biology and many other fields. Many books are written about the Golden Ratio; one such book is *The Glorious Golden Ratio*, by A. S. Posamentier and I. Lehmann (Prometheus Books, 2012).

To determine the numerical value of the Golden Ratio, $\dfrac{L}{S}$, we will change the equation $\dfrac{L}{S} = \dfrac{L+S}{L}$ to read $\dfrac{L}{S} = \dfrac{L}{L} + \dfrac{S}{L}$. We then let $x = \dfrac{L}{S}$ so that we get: $x = 1 + \dfrac{1}{x}$.

We can now solve this equation for $x$ using the quadratic formula,[1] which you may recall from high school. We then obtain the numerical value of the Golden Ratio: $\dfrac{L}{S} = x = \dfrac{1+\sqrt{5}}{2}$, which is commonly denoted by the Greek letter, phi: $\phi$.[2]

$$\phi = \frac{L}{S} = \frac{1+\sqrt{5}}{2} \approx \frac{1 + 2.2360679774997896964091736687312762354 40}{2}$$

$$\approx \frac{3.2360679774997896964091736687312762354 40}{2}$$

$$\approx 1.6180339887498948482045868343656381177 20$$

**Equation 1**

Notice what happens when we take the reciprocal of $\frac{L}{S}$, namely, $\frac{S}{L} = \frac{1}{\phi}$, so that $\frac{1}{\phi} = \frac{2}{1+\sqrt{5}}$, which, when we multiply by 1 in the form of $\frac{1-\sqrt{5}}{1-\sqrt{5}}$, we get:

$$\frac{2}{1+\sqrt{5}} \times \frac{1-\sqrt{5}}{1-\sqrt{5}} = \frac{2(1-\sqrt{5})}{1-5} = \frac{2(1-\sqrt{5})}{-4} = \frac{1-\sqrt{5}}{-2} = \frac{\sqrt{5}-1}{2} \approx$$

0.6180339887498948482045868343656381177 20.

Notice a very unusual relationship. The value of $\phi$ and $\frac{1}{\phi}$ differ by 1. That is, $\phi - \frac{1}{\phi} = 1$. From the normal relationship of reciprocals, the product of $\phi$ and $\frac{1}{\phi}$ is also equal to 1, that is, $\phi \cdot \frac{1}{\phi} = 1$. Therefore, we have two special numbers $\phi$ and $\frac{1}{\phi}$ whose difference and product is 1 — these are the only two numbers for which this is true! By the way, you might have noticed that their sum $\phi + \frac{1}{\phi} = \sqrt{5}$ because $\frac{\sqrt{5}+1}{2} + \frac{\sqrt{5}-1}{2} = \sqrt{5}$.

One reason why this ratio has captured the name "Golden" could be because of the beautiful rectangle whose ratio of sides form the Golden Ratio.

To get a more complete feeling about the *Golden Rectangle*, we offer a brief construction, one of the many ways it can be made. We begin with a unit square[3] *ABCD*, with midpoint *M* of side *AB*, and then with center at *M* we draw a circular arc *MC*, cutting the extension of side *AB* at point *E*, as shown in Figure 5. We now can claim that the line segment *AE* is partitioned into the Golden Ratio at point *B*.

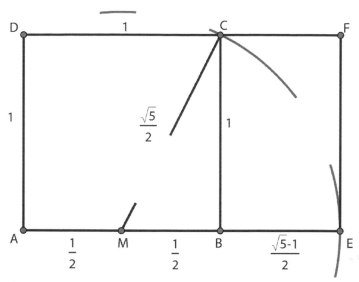

**Figure 5**

To verify this, we would have to apply the definition of the Golden Ratio: $\dfrac{AB}{BE} = \dfrac{AE}{AB}$, and see if it, in fact, holds true. Substituting the values obtained by applying the Pythagorean Theorem to $\triangle MBC$ as shown in Figure 5, we get the following:

$$MC^2 = MB^2 + BC^2 = \left(\frac{1}{2}\right)^2 + 1^2 = \frac{1}{4} + 1 = \frac{5}{4}; \text{ therefore } MC = \frac{\sqrt{5}}{2}.$$

It follows that

$$BE = ME - MB = MC - MB = \frac{\sqrt{5}}{2} - \frac{1}{2} = \frac{\sqrt{5} - 1}{2} \text{ and}$$

$$AE = AB + BE = 1 + \frac{\sqrt{5} - 1}{2} = \frac{2}{2} + \frac{\sqrt{5} - 1}{2} = \frac{\sqrt{5} + 1}{2}. \text{ We then can}$$

find the value of $\dfrac{AB}{BE} = \dfrac{AE}{AB}$, that is $\dfrac{1}{\dfrac{\sqrt{5} - 1}{2}} = \dfrac{\dfrac{\sqrt{5} + 1}{2}}{1}$, which turns out to be a true proportion because the cross products are equal. That is,

$$\left(\frac{\sqrt{5} - 1}{2}\right)\cdot\left(\frac{\sqrt{5} + 1}{2}\right) = 1 \times 1 = 1.$$

We can also see from Figure 3 that point $B$ can be said to divide the line segment $AE$ into an inner Golden Ratio because $\dfrac{AB}{AE} = \dfrac{1}{\dfrac{\sqrt{5} - 1}{2}} = \dfrac{\sqrt{5} + 1}{2} = \phi.$

Meanwhile, point $E$ divides the line segment $AB$ into an outer Golden

Ratio because $\dfrac{AE}{AB} = \dfrac{1 + \dfrac{\sqrt{5} - 1}{2}}{1} = \dfrac{\sqrt{5} + 1}{2} = \phi.$

Notice the shape of the rectangle $AEFD$ in Figure 5. The ratio of the

length to the width is the Golden Ratio: $\dfrac{AE}{EF} = \dfrac{\dfrac{\sqrt{5} + 1}{2}}{1} = \dfrac{\sqrt{5} + 1}{2} = \phi.$

Largely because of the importance of this ubiquitous number, 1.618...,

we have taken the liberty of exploring its background appropriately.


## THE NUMBER 2


The number 2 is perhaps the most common number in our number system. We use the concept of 2 when we speak of twins, duality, doubles, pairs, and so forth. The human body consists of two eyes, two ears, two arms, two legs, and so forth. The humans are separated into two sexes. Days are partitioned into two periods: night and day. In mathematics we have binary operations: addition, subtraction, multiplication, and division. Because we know that the product of two negative numbers is a positive number, we also know that two "wrongs" do not make a "right." We can also determine, by simple observation, if a given number is divisible by 2 by simply inspecting the units digit to see if it is even or odd.

The number 2 also has an important status in the following ways:

- The number 2 is the only number whose self-sum and self-product is the same, that is, $2 + 2 = 4$, and $2 \times 2 = 4$.
- The number 2 is the first prime number because it has exactly 2 divisors: 1 and 2.
- The number 2 is the only even prime number.
- The number 2 is the only prime followed consecutively by another prime number, namely, 3.
- Powers of 2 appear more frequently in mathematics than powers of any other numbers.
- Positive integers can be expressed as the sum of 2 or more consecutive integers, but only if they are not a power of 2.
- The number 2 is the only prime number whose cube is a sum of two consecutive prime numbers, that is, $2^3 = 8 = 3 + 5$.
- The number 2 is the only prime number that is not the difference of two square numbers.
- The number 2 is the only prime number that can be represented in the form $n^n + n$; which is true when $n = 1$.

- The number 2 is the smallest *deficient number*. A deficient number is one in which the sum of its divisors is less than the number itself. For example, the number 27 is a deficient number because its divisors: 1, 3, and 9 have a sum of 13 which is less than 27, thereby designating 27 as a deficient number.

The number 2 appears in Euler's famous formula regarding aspects of polyhedron, namely, vertices + faces − edges = 2.

Every number can be expressed in a unique way as the sum of powers of 2, as for example, $59 = 32 + 16 + 8 + 2 + 1$, which allows representing the number 59 in binary form as $111011_2$. In the Western world, the binary system was originated by Gottfried Wilhelm Leibniz (1646–1716) and became an essential part of Charles Babbage's Analytical Engine, which was the forerunner of today's computers.

The famous French mathematician Pierre de Fermat (1607–1665) discovered that if $n$ is a prime number then $n$ will always be a factor of $2^n − 2$. For example, when $n = 5$ then $2^5 − 2 = 30$, of which 5 is a factor. Also, when $n$ is a prime number, then $2^n − 1$ will always be a prime number as well. For example, when $n = 7$, we have $2^7 − 1 = 127$, which is a prime number.

Fermat's famous "last theorem" states that the equation $x^n + y^n = z^n$ has integer solutions, if and only if, $n$ is not greater than 2. This of course credits the Pythagorean theorem: $x^2 + y^2 = z^2$. This was finally approved correct by the British mathematician Andrew Wiles (1953–) in 1994.

The number 2 also plays a significant role in the famous Goldbach's conjecture that states that every even number greater than 2 is the sum of two prime numbers. For example, $36 = 5 + 31 = 7 + 29 = 13 + 24 = 17 + 19$. This conjecture, although never proved to be wrong, has never been approved for all cases.

The number 2 is the only number where the sum of the reciprocals of its powers equals the number itself, as we can see with the following:

$$\sum_{n=0}^{\infty} \frac{1}{2^n} = 1 + \frac{1}{2} + \frac{1}{4} + \frac{1}{8} + \frac{1}{16} + \frac{1}{32} + \cdots = 2$$

In a right triangle, the measures of whose angles are 30°, 60°, and 90°, the length of the hypotenuse is two units longer than the length of the shorter leg.

Here is a strange curiosity that was discovered recently by the British mathematician Phil Carmody (1951–) who calculated that $2^{168}$ is the largest power of 2 whose decimal expansion does not contain the digit 2. That is, $2^{168} = 74{,}144{,}419{,}156{,}711{,}147{,}060{,}143{,}317{,}175{,}368{,}453{,}031{,}918{,}731{,}001{,}856$.

Thus, we can appreciate the special role that the number 2 has in both mathematics and beyond.


## THE NUMBER 3


Because we live in a three-dimensional world, number 3 plays a significant role in determining these dimensions: in simple form, width, length and depth. When we tell a story, there are often three parts: a beginning, a middle, and an end. In mythology, such as the Norse legends, there are three worlds: the underworld, the earth, and the heavens. This is portrayed very clearly in the German composer Richard Wagner's (1813–1883) "Der Ring des Nibelungen," which is composed of a prelude (Das Rheingold), followed by three operas (Die Walküre, Siegfried, and Die Götterdämmerung). In biology we find octopuses have three hearts.

In mathematics, the number 3 plays a most significant role:

- A plane surface is defined by three noncollinear points.
- A triangle is defined by three noncollinear points as well.
- A triangle has three vertices and three sides.
- A unique circle is determined by three noncollinear points.
- The number 3 is the fourth member of the Fibonacci sequence.
- The number 3 is the only number that is equal to the sum of all the preceding positive integers: $1 + 2 = 3$.
- One of the three famous problems of antiquity was to try to divide a given angle into three equal parts using only a straight-edge and a pair of compasses.
- The famous German mathematician Carl Friedrich Gauss (1777–1855) showed that every integer is the sum of at most three triangular numbers. (Triangular numbers are those that resemble points that can be arranged in triangular form, such as the numbers 1, 3, 6, 10, 15, 21, 28, ....) For example, $39 = 28 + 10 + 1$.
- The number 3 is the only Fibonacci prime number that is also a triangular number.
- Among the set of odd numbers there will never be one that cannot be expressed as the sum of 3 or fewer prime numbers.
- Numbers that are not in the form of $4^k(8n + 7)$ can be expressed as the sum of 3 squares.
- The number 3 is the only prime number, which when 2 is added to its square, yields a prime number.

Here are some curious expressions ways using the number 3:

- $2^3 + 3 = 11$, and $2^3 - 3 = 5$
- $3^3 = 3^2 + 3^2 + 3^2 = 27$
- $3^3 + 4^3 + 5^3 = 6^3 = 216$
- $3^5 = 41 + 43 + 47 + 53 + 59$, which are five consecutive prime numbers.

Each of the following powers of 3 results in a number, where the sum of the digits is 9 or a multiple of 9:

- $3^2 = 9$, $3^3 = 27$, $3^4 = 81$ and $3^5 = 243$. The next 3 powers of 3 each has a digit-sum of 18:
- $3^6 = 729$, $3^7 = 2187$, $3^8 = 6561$. This pattern continues with digit sums that are multiples of 9.

The number 3 is the only prime number $n$ where $\dfrac{n^5 - 1}{n - 1}$ results in the square of a prime number as we see with $\dfrac{3^5 - 1}{3 - 1} = 121$, which is $11^2$.

Among the many rules for testing numbers for the divisibility, the number 3 has a particularly useful technique for testing for its divisibility. To determine if a given number is divisible by 3, one needs only to check if the sum of the digits of that given number is divisible by 3. If it is, then the number itself is divisible by 3. For example, consider the number 3,618. Because the sum of the digits: $3 + 6 + 1 + 8 = 18$, which is divisible by 3, we know that the original number 3,618 is divisible by 3.

The number 3 also plays a role in the field of magic squares because the smallest magic square is one that is $3 \times 3$ in rows and columns. Magic squares are a square arrangement of numbers with the sum of each column, row, and diagonal is the same. Following is the development of a $3 \times 3$ magic square:

The magic square we are about to construct will contain the numbers from 1 to 9 and have a sum of 45 which would have to be divided by 3 so that each row as one-third of the sum, namely, 15. We can begin by placing the middle number of the sequence from 1 to 9, which is 5 in the center position. Then we place the number 1 in a corner, as shown in Figure 6. Therefore, the

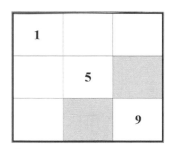

**Figure 6**

lower right-hand corner must have the 9, so that the diagonal adds up to 15, because it must have the same sum as any role or column.

Next, we notice that 2, 3, and 4 cannot be in the same row (or column) as 1 because there is no natural number less than 9, which would be large enough to occupy the third position of such a row (or column). This would leave only the two shaded positions in Figure 6 to accommodate these three numbers (2, 3, and 4). Because this cannot be the case, our first attempt was a failure: the numbers 1 and 9 may occupy only the middle positions of a row (or column).

Therefore, we have to start with one of the four possible positions remaining for 1, which are the center cell of a row or column, as we show in the first square of Figure 7. We note that the number 3 cannot be in the same row (or column) as 9, for the third number in such a row (or column) would again have to be 3, to obtain the required sum of 15. This is not possible, because a number can be used only once in the magic square. Additionally, we have seen earlier that 3 cannot be in the same row (or column) as 1. This leaves only the two shaded positions in Figure 7 for the number 3. The number opposite to 3 is always 7, because then $3 + 5 + 7 = 15$.

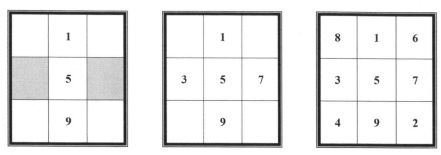

**Figure 7**

We continue with the second square in Figure 7, showing one of two possibilities for the placement of 3 and 7 (the other possibility has 3 and 7 exchanged). It is now easy to fill in the remaining numbers. There is only one such possibility, shown in the third square of Figure 7.

How many different $3 \times 3$ magic squares are there? We could start by putting the number 1 in any of the four positions in the middle of a side row or column. Then have two possibilities for placing 3. After that, the construction is unique. This produces the eight magic squares shown in Figure 8.

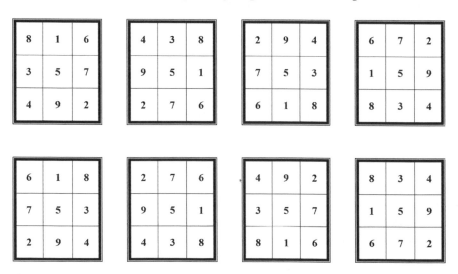

**Figure 8**

These are just some of the appearances of the first odd prime number in our number system.

## THE NUMBER 3.141592653589793238462643383279 5.... THE NUMBER KNOWN AS π.

One of the most popular numbers that folks recall from their school days is the number 3.14, which is an approximation of π. When used designate a date in the United States it would indicate March 14, which just happens to be Albert Einstein's birthday. However, it made its fame from its definition and is perhaps one of the most famous numbers in mathematics and usually referred to in shorter form simply as 3.14 and is *the ratio of the circumference of a circle to its diameter.* The symbol π was chosen to represent this ratio as $\pi = \dfrac{C}{d}$, where $C$ represents the length of the circumference and $d$ represents the length of the diameter, in 1706 by the Welsh mathematician William Jones (1675–1749). With the help of today's computers, the value of has now been found to several trillion places.

The first 100 places of π are: 3.1415926535 8979323846 2643383279 5028841971 6939937510 5820974944 5923078164 0628620899 8628034825 3421170679....

As evidence of its fame, the number π is exhibited in the most unexpected places. For example, there is a curious situation in which we can see the enthusiasm for π demonstrated for all to see. In 1937, in Hall 31 of the Palais de la Decouverte, today a Paris science museum (on Franklin D. Roosevelt Avenue), the value of π was produced with large wooden numerals on the ceiling (a cupola) in the form of a spiral. This was a nice dedication to this famous number, but there was an error: They used the 707-place approximation generated in 1874 by William Shanks, which had an error in the 528th decimal place. This was detected in 1946 and corrected on the museum's ceiling in 1949 (see Figure 9).

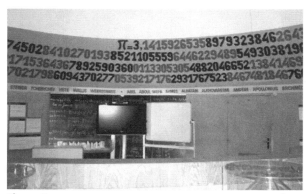

**Figure 9**

Also, in a newly renovated underground subway concourse, the Opernpassage in Vienna, Austria exhibits the value of $\pi$ to thousands of places along a mirrored wall as seen in Figure 10.

**Figure 10**

Many schools in recent years tend to celebrate March 14 as $\pi$-day, which sometimes is highlighted with a poster such as the one shown in Figure 11.

**Figure 11**

From the earlier definition of $\pi$, we can get to the very familiar formula for the circumference of a circle. The diameter, $d$, of a circle is twice the length of the radius, $d = 2r$, where $r$ is the length of the radius. If we substitute $2r$ for $d$ in our previous definition of $\pi$, we get $\pi = \dfrac{C}{2r}$, which leads us to the famous formula for the circumference of a circle: $C = 2\pi r$, an alternative of which is $C = \pi d$.

Because we conveniently recalled generating the circumference of the circle with the familiar formula $C = 2\pi r$, it is only fitting that we do the same for the popular formula for the area of a circle $\left(A = \pi r^2\right)$ with radius $r$. We begin by drawing a convenient size circle on a piece of cardboard. Divide the circle (which consists of 360°) into 16 equal arcs. This may be done by marking off consecutive arcs of 22.5° or by consecutively dividing the circle into two parts, then four parts, then bisecting each of these quarter arcs, and so on.

The 16 sectors we have constructed in Figure 12 are then to be cut apart and placed in the manner shown in the Figure 13.

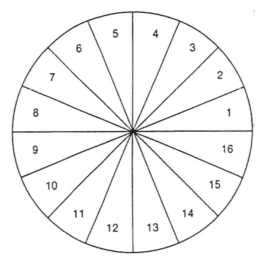

**Figure 12**

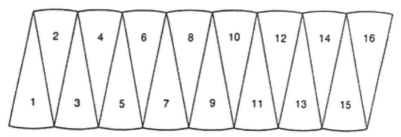

**Figure 13**

This placement suggests that we have a figure that approximates a parallelogram. That is, were the circle cut into more sectors, then the figure would look even more like a true parallelogram. Let us assume it is a parallelogram. In this case, the base would have a length of half the circumference of the original circle because half of the circle's arcs are used for each of the two sides of the approximate parallelogram. In other words, we formed something that resembles a parallelogram, where one pair of opposite sides are not straight lines, rather they are circle arcs. We will progress as though they were straight lines, realizing that we will have lost some accuracy in the process. Therefore, the length of the base is $\frac{1}{2}C$. Because $C = 2\pi r$, the base length is, therefore, $\pi r$. The area of a parallelogram is equal to the product of its base and altitude. Here the altitude is the radius $r$ of the original circle. Therefore, the area of the "parallelogram" (which is the area of the circle we just cut apart) is $(\pi r)(r) = \pi r^2$, which is the commonly known formula for the area of a circle.

There are many peculiarities in this list of digits. The British Mathematician John H. Conway (1937–2020) has indicated that if you separate the decimal value of $\pi$ into groups of 10 places, the probability of each of the 10 digits appearing in any of these blocks is about 1 in 40,000. Yet, he shows that it does occur in the seventh such group of 10 places, as you can see from the grouping in the following text:

$\pi = 3.1415926535\ 8979323846\ 2643383279\ 5028841971\ 6939937510$
5820974944 5923078164 0628620899 8628034825 3421170679 8214808651
3282306647 0938446095 5058223172 5359408128 . . . . Another way of saying this is that every other grouping of 10 digits has at least one repeating digit.

The sums of the digits of $\pi$ also show some nice results: for example, the sum of the first 144 digits in the decimal expansion is 666, a number with some curious properties as we shall see later.

On occasion, we stumble upon phenomena involving $\pi$ that have nothing whatsoever to do with a circle. For example, the probability that a randomly selected integer has only unique prime divisors[4] is $\frac{6}{\pi^2}$. Clearly this relationship has nothing to do with a circle, yet it involves the circle's ratio, $\pi$. This is just another feature that adds to the centuries-old fascination with $\pi$.

There is much to be said for the adventures of calculating the value of $\pi$. It is interesting to note that Archimedes of Syracuse (287–212 B.C.E.) showed the value of $\pi$ to lie between $3\frac{10}{71}$ and $3\frac{1}{7}$. That is,

$$3\frac{10}{71} < \pi < 3\frac{1}{7}$$
$$\frac{223}{71} < \pi < \frac{22}{7}$$
$$3.14\,08 < \pi < 3.14\,28$$

The German mathematician Ludolph van Ceulen (1540–1610) calculated $\pi$ to 35 places, so for a time the ratio $\pi$ was called *Ludolph's number*. When Ludolph van Ceulen finished his calculations he wrote the following: "Die lust heeft, can naerder comen" ("The one who has the desire, can come closer").

Another early technique for calculating $\pi$ was discovered by the English mathematician John Wallis (1616–1703), a professor of mathematics at Cambridge and Oxford universities, who subsequently published it in his book, *Arithmetica infinitorum* (1655). There he presented a formula for $\pi$ ($\frac{\pi}{2}$, which we then merely double to get $\pi$). The following is Wallis's formula:

$$\frac{\pi}{2} = \frac{2 \times 2}{1 \times 3} \times \frac{4 \times 4}{3 \times 5} \times \frac{6 \times 6}{5 \times 7} \times \frac{8 \times 8}{7 \times 9} \times \cdots \times \frac{2n \times 2n}{(2n-1)(2n+1)} \times \cdots$$

This product converges to the value of $\frac{\pi}{2}$. That means it gets closer and closer to the value of $\frac{\pi}{2}$ as the number of terms increases.

What is it about the value of $\pi$ that evokes so much fascinating? For one, it cannot be calculated by a combination of the operations of addition, subtraction, multiplication, and division, which was suspected by Aristotle (384–322 B.C.). He hypothesized that $\pi$ is an irrational number;[5] in other words, the circumference and radius of a circle are incommensurable. That means there doesn't exist a common unit of measure that will allow us to measure both the circumference and radius. This was proved in 1806[6] by the French mathematician Adrien-Marie Legendre (1752–1833)—more than two millennia later! But even more fascinating is fact that $\pi$ cannot be calculated by a combination of the operations of addition, subtraction, multiplication, division, *and square root extraction*.

This means that $\pi$ is a type of nonrational number called a transcendental number[7] by the Swiss mathematician Leonhard Euler[8] (1707–1783), but it was first proved in 1882 by the German mathematician (Carl Louis) Ferdinand Lindemann (1852–1939). Remember, it is sometimes more difficult to prove that something cannot be done than to prove it is possible to be done. Thus, for Lindemann to establish that $\pi$ could not be produced by a combination of the five operations: addition, subtraction, multiplication, division, and square root extraction, was quite an important contribution to the development of our understanding of mathematics.

There is an amazing revelation of what the Bible has as the value of $\pi$. For many years virtually all the books on the history of mathematics stated that in $\pi$'s earliest manifestation in history, namely in the Old Testament of the Bible, gave the value of $\pi$ as 3. Yet recent "detective work" shows otherwise.[9]

One always relishes the notion that a hidden code can reveal long-lost secrets. Such is the case with the common interpretation of the value of $\pi$ in the Bible. There are two places in the Bible where the exact same sentence appears—identical in every way except for one word, which is spelled dif-ferently in the two citations. The description of a pool, or fountain, in King Solomon's temple is referred to in the passages that may be found in 1 *Kings* 7:23 and 2 *Chronicles* 4:2, and reads as follows:

> And he made the molten sea[10] of ten cubits from brim to brim, round in compass, and the height thereof was five cubits; and *a line* of thirty cubits did compass it round about.

The circular structure described here is said to have a circumference of 30 cubits[11] and a diameter of 10 cubits. From this we notice that the Bible has $\pi = \dfrac{30}{10} = 3$. This is obviously a very primitive approximation of $\pi$. A late-18th-century Rabbi, Elijah of Vilna[12] (1720–1797), was one of the great modern biblical scholars, who earned the title "Gaon of Vilna" (meaning brilliance of Vilna). He came up with a remarkable discovery, one that could make most history of mathematics books faulty, if they say that the Bible approximated the value of $\pi$ as 3. Elijah of Vilna noticed that the Hebrew word for "line measure" was written differently in each of the two biblical passages mentioned in the preceding text.

In 1 *Kings* 7:23 it was written as קוה, whereas in 2 *Chronicles* 4:2 it was written as קו. Elijah applied the ancient biblical analysis technique (still used by Talmudic scholars today) called gematria, where the Hebrew letters are given their appropriate numerical values according to their sequence in the Hebrew alphabet, to the two spellings of the word for "line measure" and found the following. The letter values are: ק = 100, ו = 6, and ה = 5. Therefore, the spell-ing for "line measure" in 1 *Kings* 7:23 is קוה = 5 + 6 + 100 = 111, while in 2 *Chronicles* 4:2 the spelling קו = 6 + 100 = 106. Using gematria in an accepted way, he then took the ratio of these two values: $\dfrac{111}{106} = 1.0472$ (to four decimal places), which he considered the necessary "correction factor." By multiplying the Bible's apparent value (3) of $\pi$ by this "correction factor," one gets 3.1416, which is $\pi$ correct to four decimal places! "Wow!" is a common reaction. Such accuracy is quite astonishing for ancient times.

It is interesting to see where mathematicians got the idea to represent the ratio of the circumference of a circle to its diameter with the Greek let-ter $\pi$. According to the well-known mathematics historian, Florian Cajori (1859–1930), the symbol $\pi$ was first used in mathematics by the English math-ematician William Oughtred (1575–1660) in 1652, when he referred to the ratio of the circumference of a circle to its diameter as $\dfrac{\pi}{\delta}$, where $\pi$ represented the periphery[13] of a circle and $\delta$ represented the diameter. In 1665, John Wallis used the Hebrew letter מ (mem), to equal one-quarter of the ratio of the circumference of a circle to its diameter (what, today, we would refer to as $\frac{\pi}{4}$).

In 1706, William Jones published his book, *Synopsis Palmariorum Matheseos*, where he used $\pi$ to represent the ratio of the circumference of a circle to its diameter. This is believed to have been the first time that $\pi$ was used as it is defined today. Yet, Jones's book alone would not have made the use of the Greek letter $\pi$ to represent this geometric ratio as popular as it has become today. It was the legendary Swiss mathematician, Leonhard Euler (1707–1783), often considered the most prolific writer in the history of mathematics, who is largely responsible for today's common use of $\pi$. In 1736, Euler began using $\pi$ to represent the ratio of the circumference of a circle to its diameter. But not until he used the symbol $\pi$ in 1748 in his famous book *Introductio in analysin infinitorum* did the use of $\pi$ to represent the ratio of the circumference of a circle to its diameter become widespread.

You should now know the wonders attached to this most famous number $\pi \approx 3.14\ldots$. Despite this rather thorough introduction to this famous mathematical concept, entire books have been written about $\pi$. One such is $\pi$: *A Biography of the World's Most Mysterious Number*, by A. S. Posamentier and I. Lehmann (Prometheus Books, 2004).

## THE NUMBER 4

Before we begin to admire the number 4, we should notice that that is the only number in the English language whose name has the number of letters that it represents. The word "four" has four letters! The number 4 is the first composite number, that is, a nonprime number. It is also often referred to as the first doubly even number. By that we mean that it is divisible by 2 twice. This also allows us to determine whether numbers are divisible by 4 by simply inspecting the last 2 digits of a number to determine whether the entire number is divisible by 4. For example, if we would like to verify if the number 36,186 is divisible by 4, we need only to check to see if the number formed by the last 2 digits 86 is divisible by 4. Because 86 is not divisible by 4, the number 36,168 is also not divisible by 4. However, the number 2,356 is divisible by 4 because the number $56 = 4 \times 14$.

In number theory, we have the four-square theorem, which was proved in 1770 by the Italian mathematician Joseph Louis Lagrange (1736–1813) and states that every natural number can be expressed as the sum of four integer squares. Here are a few examples to illustrate the theorem:

$$2 = 1^2 + 1^2 + 0^2 + 0^2$$

$$31 = 1^2 + 1^2 + 2^2 + 5^2$$

$$310 = 1^2 + 2^2 + 4^2 + 17^2$$

Still within the realm of number theory, the number 4 carries some unique features. For example, it is the only composite number, $n$, which is not a factor of $(n-1)!$ For example, let's consider the composite number 6, and apply it here so that we get $(6-1)! = 5! = 5 \times 4 \times 3 \times 2 \times 1 = 120 = 4 \times 30$. Whereas, if $n = 4$, we find that $(4-1)! = 3! = 3 \times 2 \times 1 = 6$, where 4, is not a factor. This is the only case where this is true.

Another unique feature of the number 4 is called the Brocard problem. This searches for numbers, where 1 greater than the factorial of a number will be a perfect square. In other words, when will the number $n! + 1$ will be a perfect square? The smallest of these is the number 4 as $4! + 1 = 25 = 5^2$. The only other two numbers that have been found to have this property are the numbers 5 and 7, which we will highlight later.

Pierre de Fermat discovered a unique feature of the number 4, which is that every prime number one greater than a multiple of 4 can be expressed uniquely as a sum of 2 squares. For example, consider the number 17, which is one greater than 16, a multiple of 4, can be expressed as $1^2 + 4^2$. Another example, the number $45(44 + 1)$ can be expressed as $3^2 + 6^2 = 45$. This can be extended in that *no* number *one less than* a multiple of 4 can be expressed as a sum of 2 squares.

Another surprising feature of the number 4 is what occurs when we take any number, and then take the sum of the squares of its digits and continuing this process with each resulting number, you will either end up with the number 4 or occasionally, by the number 1. For example, let's begin with the number 5 and continually take the sum of the squares of its digits until we notice a pattern: $5 - 25 - 29 - 85 - 89 - 145 - 42 - 20 - 4$. When we continue this with the number 4 notice what results: $4 - 16 - 37 - 58 - 89 - 145 - 42 - 20 - 4$. Thus, we see that the cycle continues. Whenever it reaches the number 4.

In geometry, one of the Platonic solids is the tetrahedron which highlights the number 4 as it has four faces, and four vertices, as can be seen in Figure 14.

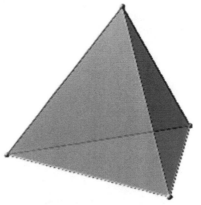

**Figure 14**

The *Four-Color Map Problem* dates back to 1852 when Francis Guthrie (1831–1899), while trying to color the map of the counties of England, noticed that four colors sufficed. He asked his brother Frederick if it was true that *any* map can be colored using only four colors in such a way that adjacent regions (i.e., those sharing a common boundary segment, not just a point) receive different colors. Frederick Guthrie then communicated the conjecture to the British mathematician, Augustus de Morgan (1806–1871). As early as 1879, the British mathematician, Alfred B. Kempe (1849–1922), produced an attempted proof of this conjecture, but in 1890 it was shown to be wrong by the British mathematician Percy J. Heawood (1861–1955). Many other subsequent attempts also proved fallacious. Not until 1976 was this so-called four-color map problem solved by two mathematicians, Kenneth Appel (1932–2013) and Wolfgang Haken (1928–), who, using a computer, considered all possible maps and established that it was never necessary to use more than four colors to color a map so that no two territories, sharing a common border, would be represented by the same color.[14] They used an IBM 360 that required about 1,200 hours to test the 1,936 cases, which later turned out to require inspecting another 1,476 cases. This "computer proof" was not widely accepted by pure mathematicians. Yet, in 2004, the mathematicians Benjamin Werner und Georges Gonthier produced a formal mathematical solution to the four-color map problem that validated Appel's and Hacken's assertion. The attractive aspect of the four-color map problem lies in the fact that it is very easily understood, but the solution has proved to be very elusive and very complex.

Using the number 4, there are some curious ways of generating a prime number as follows:

$4!^4 + 1 = 331,777$, which is a prime number.

$1!^1 + 2!^2 + 3!^3 + 4!^4 = 331,997$, which is a prime number.

$4^4 - 4! + 1 = 233$, which is a prime number.

$4444^4 + 1 = 390,028,372,746,497$, which is a prime number.

$4444^{4 \times 4} + 1 = 231,411,429,064,878,465,277,548,197,361,116,672,093,$
$$463,781,709,982,52,896,257, \text{which is a prime number.}$$

There are only four ways to create prime numbers with four digits, whose sum is equal to 4. They are the following 1,201, 1,021, 2,011, and 3,001. There are only four ways to create four-digit prime numbers using the first four natural numbers, which are 1,423, 2,143, 2,341 and 4,231.

Merely as recreational, we can use the number 4 to generate other numbers as you can see with the following examples up to 100:

$0 = 44 - 44 = \dfrac{4}{4} - \dfrac{4}{4}$

$1 = \dfrac{4+4}{4+4} = \dfrac{\sqrt{44}}{\sqrt{44}} = \dfrac{4+4-4}{4}$

$2 = \dfrac{4 \bullet 4}{4+4} = \dfrac{4-4}{4} + \sqrt{4} = \dfrac{4}{4} + \dfrac{4}{4}$

$3 = \dfrac{4+4+4}{4} = \sqrt{4} + \sqrt{4} - \dfrac{4}{4} = \dfrac{4 \cdot 4 - 4}{4} = 4 - 4^{4-4}$

$4 = \dfrac{4-4}{4} + 4 = \dfrac{\sqrt{4 \bullet 4 \bullet 4}}{4} = (4-4) \cdot 4 + 4$

$5 = \dfrac{4 \bullet 4 + 4}{4}$

$6 = \dfrac{4+4}{4} + 4 = \dfrac{4\sqrt{4}}{4} + 4$

$7 = \dfrac{44}{4} - 4 = \sqrt{4} + 4 + \dfrac{4}{4} = (4+4) - \dfrac{4}{4}$

$8 = 4 \bullet 4 - 4 - 4 = \dfrac{4(4+4)}{4} = 4+4+4-4$

$9 = \dfrac{44}{4} - \sqrt{4} = 4\sqrt{4} + \dfrac{4}{4} = \dfrac{4}{4} + 4 + 4$

$10 = 4+4+4 - \sqrt{4} = \dfrac{44-4}{4}$

$11 = \dfrac{4}{4} + \dfrac{4}{.4} = \dfrac{44}{\sqrt{4}+\sqrt{4}}$

$12 = \dfrac{4 \bullet 4}{\sqrt{4}} + 4 = 4 \bullet 4 - \sqrt{4} - \sqrt{4} = \dfrac{44+4}{4}$

$13 = \dfrac{44}{4} + \sqrt{4}$

$14 = 4 \bullet 4 - 4 + \sqrt{4} = 4+4+4+\sqrt{4} = \dfrac{4!}{4+4+4} = 4! - (4+4+\sqrt{4})$

$15 = \dfrac{44}{4} + 4 = \dfrac{\sqrt{4}+\sqrt{4}+\sqrt{4}}{.4}$

$16 = 4 \bullet 4 - 4 + 4 = \dfrac{4 \bullet 4 \bullet 4}{4} = 4+4+4+4$

$17 = 4 \bullet 4 + \dfrac{4}{4}$

$18 = \dfrac{44}{\sqrt{4}} - 4 = 4 \bullet 4 + 4 - \sqrt{4} = 4 \bullet 4 + \dfrac{4}{\sqrt{4}} = \dfrac{4!+4!+4!}{4}$

$19 = \dfrac{4+\sqrt{4}}{.4} + 4 = 4! - 4 - \dfrac{4}{4}$

$20 = 4 \bullet 4 + \sqrt{4} + \sqrt{4} = \left(4 + \dfrac{4}{4}\right) \cdot 4$

$21 = 4! - 4 + \dfrac{4}{4}$

$22 = \dfrac{4}{4}(4!) - \sqrt{4} = 4! - \dfrac{((4+4)/4)}{4} = \dfrac{44}{4} \cdot \sqrt{4} = -4 + \dfrac{4}{4}$

$23 = 4! - \sqrt{4} + \dfrac{4}{4} = 4! - 4^{4-4}$

$24 = 4 \cdot 4 + 4 + 4$

$25 = 4! - \sqrt{4} + \dfrac{4}{4} = 4! + \sqrt{(4+4-4)} = \left(4 + \dfrac{4}{4}\right)^{\sqrt{4}}$

$26 = \dfrac{4}{4}(4!) + \sqrt{4} = 4! + \sqrt{4+4-4} = 4 + \dfrac{44}{\sqrt{4}}$

$27 = 4! + 4 - \dfrac{4}{4}$

$28 = (4+4) \cdot 4 - 4 = 44 - 4 \cdot 4$

$29 = 4! + 4 + \dfrac{4}{4}$

$30 = 4! + 4 + 4 - \sqrt{4}$

$31 = \dfrac{((4 + \sqrt{4})! + 4!)}{4!}$

$32 = (4 \cdot 4) + (4 \cdot 4)$

$33 = 4! + 4 + \dfrac{\sqrt{4}}{.4}$

$34 = \left(4 \cdot 4 \cdot \sqrt{4}\right) + \sqrt{4} = 4! + \left(\dfrac{4!}{4}\right) + 4 = \sqrt{(4^4)} \cdot \sqrt{4} + \sqrt{4}$

$35 = 4! + \dfrac{44}{4}$

$36 = (4+4) \cdot 4 + 4 = 44 - 4 - 4$

$37 = 4! + \dfrac{\left(4! + \sqrt{4}\right)}{\sqrt{4}}$

$38 = 44 - \dfrac{4!}{4}$

$39 = 4! + \dfrac{4!}{4 \cdot .4}$

$40 = (4! - 4) + (4! - 4) = 4 \cdot \left(4 + 4 + \sqrt{4}\right)$

$41 = \dfrac{4! + \sqrt{4}}{.4} - 4!$

$42 = 44 - 4 + \sqrt{4} = (4! + 4!) - \dfrac{4!}{4}$

$43 = 44 - \left(\dfrac{4}{4}\right)$

$44 = 44 + 4 - 4$

$45 = 44 + \dfrac{4}{4}$

$46 = 44 + 4 - \sqrt{4} = (4! + 4!) - \left(\dfrac{4}{\sqrt{4}}\right)$

$47 = (4! + 4!) - \dfrac{4}{4}$

$48 = (4 \cdot 4 - 4) \cdot 4 = 4 \cdot (4 + 4 + 4)$

$49 = (4! + 4!) + \dfrac{4}{4}$

$50 = 44 + \left(\dfrac{4!}{4}\right) = 44 + 4 + \sqrt{4}$

$51 = \dfrac{(4! - 4 + .4)}{.4}$

$52 = 44 + 4 + 4$

$53 = 4! + 4! + \dfrac{\sqrt{4}}{.4}$

$54 = 4! + 4! + 4 + \sqrt{4}$

$55 = \dfrac{(4! - 4 + \sqrt{4})}{.4}$

$56 = 4! + 4! + 4 + 4 = 4 \cdot (4 \cdot 4 - \sqrt{4})$

$57 = \left(\dfrac{4! - \sqrt{4}}{.4}\right) + \sqrt{4}$

$58 = ((4! + 4) \cdot \sqrt{4}) + \sqrt{4} = 4! + 4! + \dfrac{4}{.4}$

$59 = \dfrac{(4! - \sqrt{4})}{.4} + 4 = \dfrac{4!}{.4} - \dfrac{4}{4}$

$60 = 4 \cdot 4 \cdot 4 - 4 = \dfrac{4^4}{4} - 4 = 44 + 4 \cdot 4$

$61 = \dfrac{(4! + \sqrt{4})}{.4} - 4 = \dfrac{4!}{.4} + \dfrac{4}{4}$

$62 = 4 \cdot 4 \cdot 4 - \sqrt{4}$

$63 = \dfrac{(4^4 - 4)}{4}$

$64 = 4\sqrt{4} \cdot 4\sqrt{4} = 4 \cdot (4! - 4 - 4) = (4 + 4) \cdot (4 + 4)$

$65 = \dfrac{(4^4 + 4)}{4}$

$66 = 4 \cdot 4 \cdot 4 + \sqrt{4}$

$67 = \dfrac{4! + \sqrt{4}}{.4} + \sqrt{4}$

$68 = 4 \cdot 4 \cdot 4 + 4 = \dfrac{4^4}{4} + 4$

$69 = \left(\dfrac{4! + \sqrt{4}}{.4}\right) + 4$

$70 = \dfrac{(4 + 4)!}{4! \cdot 4!} = 44 + 4! + \sqrt{4}$

$71 = \dfrac{4! + 4.4}{.4}$

$72 = 44 + 4! + 4 = 4 \cdot \left(4 \cdot 4 + \sqrt{4}\right)$

$73 = \dfrac{4! \sqrt{4} + \sqrt{.4}}{\sqrt{.4}}$

$74 = 4! + 4! + 4! + \sqrt{4}$

$75 = \dfrac{4! + 4 + \sqrt{4}}{.4}$

$76 = (4! + 4! + 4!) + 4$

$77 = \left(\dfrac{4}{.4}\right)^{\sqrt{4}} - 4$

$78 = 4 \cdot (4! - 4) - \sqrt{4}$

$79 = 4! + \dfrac{4! - \sqrt{4}}{.4}$

$80 = (4 \cdot 4 + 4) \cdot 4$

$81 = \left(4 - \left(\dfrac{4}{4}\right)\right)^4 = \left(\dfrac{4!}{4\sqrt{4}}\right)^4$

$82 = 4 \cdot (4! - 4) + \sqrt{4}$

$83 = \dfrac{4! - .4}{.4} + 4!$

$84 = 44 \cdot \sqrt{4} - 4$

$85 = \dfrac{\dfrac{4! + 4}{.4}}{.4}$

$86 = 44 \cdot \sqrt{4} - \sqrt{4}$

$87 = 4 \cdot 4! - \dfrac{4}{.4} = 44\sqrt{4} - i^4$

$88 = 4 \cdot 4! - 4 - 4 = 44 + 44$

$89 = 4! + \dfrac{4! + \sqrt{4}}{.4}$

$$90 = 4 \cdot 4! - 4 - \sqrt{4} = 44 \cdot \sqrt{4} + \sqrt{4}$$

$$91 = 4 \cdot 4! - \frac{-\sqrt{4}}{.4}$$

$$92 = 4 \cdot 4! - \sqrt{4} - \sqrt{4} = 44 \cdot \sqrt{4} + 4$$

$$93 = 4 \cdot 4! - \frac{4}{.4}$$

$$94 = 4 \cdot 4! + \sqrt{4} - 4$$

$$95 = 4 \cdot 4! - \frac{4}{4}$$

$$96 = 4 \cdot 4! + 4 - 4 = 4! + 4! + 4! + 4!$$

$$97 = 4 \cdot 4! + \frac{4}{4}$$

$$98 = 4 \cdot 4! + 4 - \sqrt{4}$$

$$99 = \frac{44}{.44} = 4 \cdot 4! + \frac{\sqrt{4}}{\sqrt{.04}} = \frac{4}{4\%} - \frac{4}{4}$$

$$100 = 4 \cdot 4! + \sqrt{4} + \sqrt{4} = \left(\frac{4}{.4}\right) \cdot \left(\frac{4}{.4}\right) = \frac{44}{.44}$$

**Equation 2**

## THE NUMBER 5

One of the earliest numbers that a child learns to grasp is the number representing the fingers on one hand. Later on, teenagers use that number when they give each other a "high-5." Britain's security service is known as the MI 5, referring to Military Intelligence, section 5. The Fifth Amendment of the US Constitution, which is often referred to in court as "pleading the 5th" absolving the defendant from self-incrimination. The US 5-cent coin, the nickel, weighs exactly 5 grams. In Judaism, the Torah consists of the five Books of Moses. In East Asian tradition there are five elements, which are earth, water, fire, wood, and metal. While in Hinduism the five elements are earth, fire, water, air, and space. Essentially, the number 5 is used in many cultures, which could, perhaps, result from the human hand—the number of fingers! The English language has five vowels: a, e, i, o, and u. When the temperature measured in Celsius is ±5 degrees it yields in Fahrenheit 41° and 23°. This is the only time when prime numbers are equivalent in both temperature measures.

In mathematics, it is clear that the number 5 is a prime number. However, it is considered a *good prime* number. One may justifiably question why the number 5 is given this special distinction? A good prime is a number who square is greater than the product of any two other primes, which, in the list of primes, are at the same number of positions before and after it. When we look at the initial list of primes: 2, 3, 5, 7, 11, we find that for the two prime numbers 3 and 7, equally spaced from the 5, we have $5^2 > 3 \times 7$ and for the primes 2 and 11, we find that $5^2 > 2 \times 11$, thus, qualifying 5 as a good prime.

The number 5 further distinguishes itself amongst the prime numbers in that it is the only prime number that is a member of two pairs of twin primes. Twin primes are two prime numbers whose difference is 2. Specifically, these two twin primes involving the number 5 are (3, 5) and (5, 7). It should be clear that 5 can be the only prime number that is the sum of two consecutive prime numbers, namely, $2 + 3 = 5$ because there can be no other consecutive prime numbers as all the rest of the prime numbers are odd numbers.

The number 5 also distinguishes itself amongst the numbers because it satisfies three characteristics. The number 5 is the sum of two primes squared $\left(1^2 + 2^2\right)$, the number 5 is also the difference of two primes squared $\left(3^2 - 2^2\right)$, and finally the number 5 is the average of another two primes squared $\left(1^2 + 3^2\right)$.

One could argue that the number 5 is one of the most prevalent numbers in American society because wherever you look at an American flag, the stars are regular pentagrams, which are five-cornered stars. The regular pentagram, which we show in Figure 15, is constructed with five lines and has the unique feature that the lines intersect each other in segments that form the golden ratio. Once again, we encounter the Golden Ratio $\dfrac{AB}{BC} = \dfrac{BC}{AC} \approx 0.618...,$ which is described in greater detail on page 6.

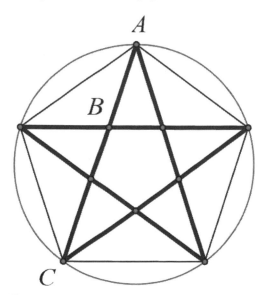

**Figure 15**

Also, when joining the five vertices of the pentagram, we formed a regular pentagon, which is a five-sided regular polygon. This shape is also prevalent as the faces of a dodecahedron, shown in Figure 16.

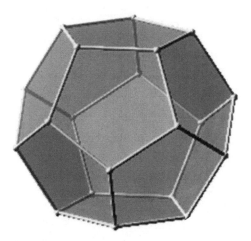

**Figure 16**

There are exactly five Platonic solids, which are tetrahedron, hexahedron (cube), octahedron, dodecahedron, and icosahedron (see Figure 17).

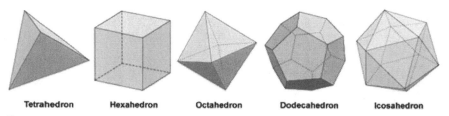

| Tetrahedron | Hexahedron | Octahedron | Dodecahedron | Icosahedron |

**Figure 17**

A few more unusual facts about the number 5 are that it can be expressed as the sum of consecutive squares as we show here: $1^2 + 2^2 = 5$, and $3^2 + 4^2 = 5^2$. Also, as we noted previously with the number 4, the number 5 factorial increased by 1 yields a square number: $5! + 1 = (5 \times 4 \times 3 \times 2 \times 1) + 1 = 121 = 11^2$, which we can recall is referred to as the Brocard problem.

Another peculiarity of the number 5 is that all powers of the number 5 have a units digit of 5, such as: $5^2 = 25$, $5^7 = 78125$, $5^{11} = 48,828,125$, and $5^{16} = 152,587,890,625$.

There are times when a number allows a strange relationship because of the numbers involved. This is the case with the number 5, where we can express an equality using only the number 5, that is, $5! + 5 = 5 \times 5 \times 5$.

Another uniqueness of the number 5 is that it is the only prime digit that can be the units digit of the perfect square. Furthermore, the number 5 appears in no other prime number as a units digit other than its original appearance as the prime number 5.

When looking for some further uniquenesses of the number 5, we find that it is the lead number of the smallest group of prime numbers in arithmetic progression, namely, 5, 11, 17, 23, 29, where the difference between these primes is 6.

As if the number 5 needed any further specialities, it is amazing how the Golden Ratio can be obtained using only the number 5. We see this with the following:

$$5^{.5} \times .5 + .5 = 1.6180339887498948....$$

Although this may be obvious, is worth pointing out that the number 5 is the only prime number who square (25) is comprised of only prime digits.

The number 5! = 120 is the largest factorial that is also a triangular number. Here are the first few triangular numbers: 0, 1, 3, 6, 10, 15, 21, 28, 36, 45, 55, 66, 78, 91, 105, **120**, 136, 153, 171, 190, 210, 231, 253, 276, 300, 325, 351, 378, 406, 435, 465, 496, 528, 561, 595, 630, 666… (for more about triangular numbers see page 12).

Perhaps this may be a bit trivial, however, when looking for unusual aspects of a number, it's placement can also be an interesting factor. Among the famous Fibonacci numbers, the number 5 is the only number equal to its place position among the Fibonacci numbers in that it is the **5th** Fibonacci number: 1, 1, 2, 3, **5**, 8, 13, 21, 34, 55, 89, …. (For more on these numbers see: A. S. Posamentier and I. Lehmann, *The (Fabulous) Fibonacci Numbers*. With Afterword by Herbert Hauptman, Nobel Laureate. (Prometheus Books, 2007.)

Beginning with the number 5, a curious pattern can generate prime numbers, as we see with the following: 5! + 4! − 3! + 2! − 1! = 139, and 5! − 4! + 3! − 2! + 1! = 101, where each of the numbers 139 and 101 are prime numbers; moreover, each of these numbers is a member of a pair of twin primes.

The Peruvian mathematician Harald Helfgott (1977–) proved that every odd number greater than 5 can be expressed as a sum of three prime numbers. For example, 9 = 2 + 2 + 5, or 21 = 5 + 5 + 11.

In the field of number theory, the number 5 tends to further distinguish itself as it joins numbers that are referred to as *Sophie Germain primes*. A prime number, $n$, which enables $2n + 1$ to also be a prime number is called a Sophie Germain prime. In the case of the number 5, we have $n = 5$, and thereby $(2 \times 5) + 1 = 11$, which is a prime number. Another example is number 11 because $(2 \times 11) + 1 = 23$. Both prime numbers 5 and 11 are referred to as Sophie Germain primes.

According to the German mathematician George Bernard Riemann (1826–1866). Every odd integer greater than 1 can be written as a sum of at most five prime numbers. In other words, you never need a sum of more than five prime numbers to represent an odd integer.

There are many more peculiarities surrounding the number 5, where it is used in book titles, movie titles, and many aspects of science. A motivated reader may wish to search for many of these.

## THE NUMBER 6

When asked where the number 6 would come up in our everyday existence, one could respond rather quickly that a cube has six faces. The number 6 is also a triangular number as it can be represented with six dots placed appropriately to form an equilateral triangle, as we can see in Figure 18. However, the number 6 also has a special property in that its square, $6^2 = 36$, is also a triangular number. Other than the number 1, the next larger triangular number that has this property as 660 digits!

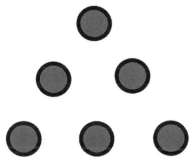

**Figure 18**

The six-pointed star, or hexagram, is formed by two overlapping equilateral triangles, and is often referred to as "The Star of David" (see Figure 19).

**Figure 19**

If one wants to be a perfectionist one could say that the sum of prime numbers up to 6 $(2 + 3 + 5 = 10)$ and the sum of composite numbers up to 6 $(4 + 6 = 10)$ is the same. Strangely enough, the number 6 is a factor of every number situated between a pair of twin prime numbers. Recall, that twin prime numbers are two prime numbers that differ by 2, such as the twin prime numbers 17 and 19 that surround the number 18 where the number 6 is one of its factors.

Playing with the number 6, we can see how it can be used to create a prime number in a rather strange fashion as follows: $6 \times 66 \times 666 \times 6,666 \times 66,666 \times 666,666 + 1 = 781,353,261,02,739,761,857$, which is a prime number. Analogously, rather than multiplying the numbers, we can also add the numbers to get $6 + 66 + 666 + 6,666 + 66,666 + 666,666 + 1 = 740,737$, which is also a prime number.

Extending this search for prime numbers using the number 6 as a generator, we can also create primes in the following curious ways: $6 + 1 = 7, 6 + 66 + 1 = 73, 6 \times 66 + 1 = 397, 6 + 66 + 666 + 1 = 739$, and $6 \times 66 \times 666 + 1 = 263,737$, which are all prime numbers.

Furthermore, beginning with the number 6 we can set up an interesting pattern that leads to a prime number, as follows: $6^6 + 5^5 + 4^4 + 3^3 + 2^2 + 1^1 = 50,069$, which is a prime number.

In mathematics, certain numbers are designated as *perfect numbers*, which are numbers the sum of whose factors is equal to the number itself. The smallest of these perfect numbers is the number 6 because its factors $1 + 2 + 3 = 6$. It is also the only perfect number located between twin prime numbers. To take this a step further, it is the only perfect number that is *not* the sum of cubes, as would be the case with the next perfect number, $28 = 1^3 + 2^3$, and the next perfect number, $496 = 3^3 + 4^3 + 4^3 + 5^3 + 6^3$. As with other perfect numbers, the sum of the reciprocals of the divisors of the number 6, namely, $\frac{1}{1} + \frac{1}{2} + \frac{1}{3} + \frac{1}{6} = 2$. While we are considering the proper divisors of the number 6, we should point out that they are the unique solution of the cubic equation: $x^3 + y^3 + z^3 = 6xyz$, that is, $x = 1, y = 2$, and $z = 3$.

Another aspect of the number 6 which makes it unique, is that it is the only number that can be expressed as the sum and product of the same three numbers, namely, $6 = 1 \times 2 \times 3 = 1 + 2 + 3$. Furthermore, $3! \times 5! = 720 = 6!$ Unexpectedly, we can also use the number 3 to represent the number 6 in the following way: $6 = \sqrt{1^3 + 2^3 + 3^3}$.

Analogous to an aspect enjoyed by the number 5, we should also take note that all powers of 6 have a units digit of 6, such as the first few powers of 6, which are: $6^6 = 6, 6^2 = 36, 6^3 = 216, 6^4 = 1,296, 6^5 = 7,776, 6^6 = 1,679,616, \ldots.$

In the realm of prime numbers, the number 6 also has a unique characteristic. Every prime number greater than 3 can be expressed in the form of $6n \pm 1$. For example, here are a few sample prime numbers expressed in this fashion:

$$37 = (6 \times 6) - 1, \ 61 = (6 \times 10) + 1, 71 = (6 \times 12) - 1, \text{ etc.}$$

The number 6 also makes a very strong presence in a rather classic theorem named for the French mathematician Blaise Pascal (1623–1662). Consider six points on a circle to be the vertices of a hexagon *ABCDEF* on a circle *c*. Then the three points of intersection of the opposite sides of the hexagon, namely, $P = AB \cap DE, Q = BC \cap EF$ and $R = CD \cap FA$, will always lie on a common line *l*. This is shown in Figure 20.

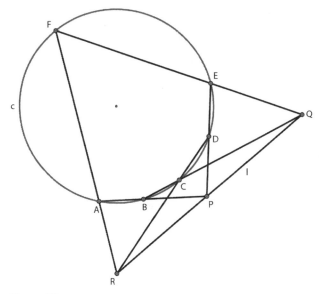

**Figure 20**

Suppose we now move the points *A*, *B*, *C*, *D*, *E*, and *F* to other positions and in another order along circle *c*, while maintaining the relationship of what were previously considered opposite sides of the hexagon. We find that the collinearity of the intersection points of the previously determined opposite sides is still true, as illustrated in Figure 21. What we have done here with six points on a circle also holds true with six points on an ellipse. Thus, the number 6 continues to fascinate us in the field of geometry.

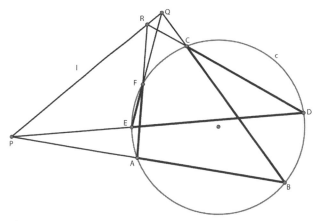

**Figure 21**

We will end our celebration of the number 6 with a cute little number trick that will always result in the number 6. We begin by taking any three consecutive numbers, where the largest is divisible by 3. We then add the numbers and then add the digits of that sum, and continue this process until single digit is reached, and you will then find that the result will always be 6. For example, to dramatize our technique, let's take three consecutive extremely large numbers and add them:

356,842,780 + 356,842,781 + 356,842,782 = 1,070,528,343. Now we find the sum of those digits: 1 + 0 + 7 + 0 + 5 + 2 + 8 + 3 + 4 + 3 = 33, and then continue to take the sum of the digits: 3 + 3 = 6, which was our desired result.

## THE NUMBER 7

What typically comes to mind first, when asked about the number 7, is the number of days in a week. And the number 7 is often considered a lucky number in Western culture. When folks are happy, they referred to themselves as being "in 7th heaven." Maximum number of games that can be played in a Major League Baseball World Series is 7. Yet it should be noted that when writing the number 7 in the United States it is written differently than in most European countries, where it is written as 7. A famous problem in mathematics that was ultimately solved in 1736 by the famous Swiss mathematician Leonhard Euler (1707–1783) involved finding a path to exactly once cross each of the seven bridges in the Prussian city of Königsberg, which eventually led to today's theory of networks.[15] Inspecting a die would reveal that the sum of the dots on opposite sides of the die is 7. The Statue of Liberty's crown has seven spikes that were intended to represent the seven seas in the world (Figure 22).

**Figure 22**

The number of minutes in exactly one week is twice 7!, which is $2 \times (1 \times 2 \times 3 \times 4 \times 5 \times 6 \times 7) = 2 \times 5040 = 10,080 = 7 \times 24 \times 60$.

There is a clever rule for determining if a given number is divisible by 7. The rule for divisibility by 7: *Delete the last digit from the given number, and then subtract twice this deleted digit from the remaining number. If the result is divisible by 7, the original number is divisible by 7. This process may be repeated if the result is too large for simple inspection for divisibility by 7.*

To see how this rule works, we will consider one example with the number 876,547 to determine if it is divisibility by 7. Beginning with 876,547 and delete its units digit, 7, and subtract its double, 14, from the remaining number: $87,654 - 14 = 87,640$. Because we cannot yet visually inspect the resulting number for divisibility by 7 we continue the process with the resulting number 87,640. Again, delete its units digit, 0, and subtract its double, still 0, from the remaining number; we get: $8,764 - 0 = 8,764$. Because this did not bring us any closer to visually determining if the resulting number is divisible by 7, delete its units digit, 4, from the number 8,764 and subtract its double, 8, from the remaining number; we get: $876 - 8 = 868$. Because we still cannot visually inspect the resulting number, 868, for divisibility by 7, we continue the process. We delete its units digit, 8, from the number 868 and subtract its double, 16, from the remaining number we get: $86 - 16 = 70$, which we can easily see is divisible by 7. Therefore, the number 876,547 is divisible by 7.

Another method for testing divisibility by 7 is as follows: take this large number and going from right to left form three-digit numbers adding and subtracting these numbers alternatively. If the resulting number is divisible by 7, then the original number was also divisible by 7. For example, consider the large number 8,641,969. Breaking up the number in groups of three from right to left and adding and subtracting alternative numbers, we get: $969 - 641 + 8 = 336$, which is divisible by 7 (as $7 \times 48 = 336$). Therefore, the original number 8,641,969 is divisible by 7.

A curious fact about the number 7 is that it is equal to the difference between the product and the sum of the previous two primes: $3 \times 5 = 15$, and $3 + 5 = 8$, and therefore $15 - 8 = 7$.

There is a curious arithmetic progression of six primes, beginning with the number 7, namely, 7, 37, 67, 97, 127, 157, where the common difference between these primes is 30. The number 7 is also the third of the Brocard's numbers as $7! + 1 = 5041 = 71^2$. It is believed that 7 is the largest prime number where its factorial +1 is a squared prime number. (Recall, the previous two numbers were 4 and 5.)

The number 7 also provides us with some arithmetic curiosities, such as when we divide 999,999 by 7, we get 142,857. At first glance, this does not appear to be anything special. However, if you consider the repeating decimal value of $\frac{1}{7} = 0.143857\ 143857\ 143857\ 143857, \dots$, which is quite spectacular. This can be taken a step further when we consider $\frac{2}{7} = 0.285714\ 285714\ 285714\ 285714\ 285714\dots$. When we consider $\frac{3}{7} = 0.428571\ 428571\ 428571\ 428571\ 428571, \dots$ We leave to the reader, the task of drawing conclusion to this surprising pattern. (Hint: we will revisit the repeating portion of the decimal 143,857 later in the book.)

Arriving at prime numbers is always interesting and so the following seven terms generate a prime number: $7! - 6! + 5! - 4! + 3! - 2! + 1! = 4{,}421$, which is the 601st prime number.

In the opposite order we can generate a pair of twin primes with the number 7 as follows:

1! × 2! × 3! × 4! × 5! × 6! × 7! = 125,411,328,000, then the twin primes are:

125,411,328,000 + 1 = 125,411,328,001,   and   125,411,328,000 − 1 = 125,411,327,999.

The sum of the seven consecutive prime numbers beginning with the number 7 is equal to 7 times the seventh prime number which is 17. That is, 7 + 11 + 13 + 17 + 19 + 23 + 29 = 119 = 7 × 17.

Can you see the amazing pattern that the number 7 produces here:
$1^7 + 4^7 + 4^7 + 5^7 + 9^7 + 9^7 + 2^7 + 9^7 = 14,459,929$?

The number 7 used seven times can produce a number the sum of whose digits is 7, as follows:
$$\frac{(77 + 777)}{7 + 7} = \frac{854}{14} = 61$$ and $6 + 1 = 7$. Furthermore, there are seven prime numbers from 7 to 77 which include a digit of 7, and they are: 7, 17, 37, 47, 67, 71, 73. Here is a cute relationship that generates the number 7: $2^5 − 5^2 = 32 − 25 = 7$.

Another peculiarity can be seen when the sum of the first seven primes (58) is divided by the seventh prime (17) the remainder is 7. Also, the sum of the primes up to 7 (which is 2 + 3 + 5 + 7 = 17) is equal to the seventh prime (17).

In geometry, the number 7 has a rather negative distinction as it is the smallest number of sides of a regular polygon, a heptagon, that cannot be constructed with a straight-edge and a pair of compasses. Yet, the United Kingdom has two heptagonal-based coins, namely £0.50 and £0.20, which is shown in Figure 23. For these coins to have the properties of the circle, circular arcs have been placed on each side, forming a Reuleaux heptagon.[16]

**Figure 23**

## THE NUMBER 8

In the Chinese culture, the number 8 is a lucky number. Folks there will pay a handsome price to get a license plate with several 8s on it. Furthermore, many airline flights to China are comprised of a number of 8s, such as Air Canada AC88, United Airlines UA888, UA88, UA8, British Airways BA88, KLM KL888, just to name a few. In 2003, the phone number in China +86 28 8888 8888 was sold to Sichuan Airlines for US$42,500. The 2008 Olympics in Beijing, China, were officially started on August 8, 2008 at 8:08 PM. Writing this symbolically, it appears as 8-8-08 at 8:08:08. The number 8 is also considered a holy number in the Japanese culture.

The number 8 also has a prominent place in chemistry because it represents the atomic number of oxygen. On August 24, 2006, when Pluto was designated a dwarf planet, it left eight planets orbiting the sun. In the biological world, we find that the number 8 has a place of significance, in that the number of tentacles of an octopus is eight, a spider has eight legs, and a human being has eight teeth in each quadrant, with the eighth tooth being the wisdom tooth. In technology, the number 8 is represented by V-8 automobile engines, and 8mm films, and Super-8 films. There were also eight-track cartridges for musical recordings. In card games, there is a game called "crazy eights" where the 8-card is a wild card, and in the game of pool the 8-ball has a significant role.

By visual inspection, it is always easy to determine if a given number is divisible by 8. All one needs to do is to inspect the last three digits of the number, and if that three-digit number is divisible by 8, then the entire number is divisible by 8.

Amongst the famous Fibonacci numbers: 1, 1, 2, 3, 5, 8, 13, 21, 34, 55,..., the number 8 is the only member of this sequence (aside from the number 1) that is a perfect cube, that is, $8 = 2^3$.... In the Fibonacci sequence no number after the number 8 will be 1 greater or 1 less than a prime number. Yet, the eighth Fibonacci number, which is 21, is 8 greater than a prime number (13) and 8 less than a prime number (29).

Aside from the number 1, the number 8 is the smallest number where the sum of the digits 8 cubed is equal to 8, that is, $8^3 = 512$, and $5 + 1 + 2 = 8$. The number 8 is also the only prime power, which is 1 less than another prime power. That can be seen by noting that $2^3 + 1 = 3^2$.

As we mentioned earlier, triangular numbers are those representing dots that can be arranged in the form of an equilateral triangle, as shown in Figure 24.

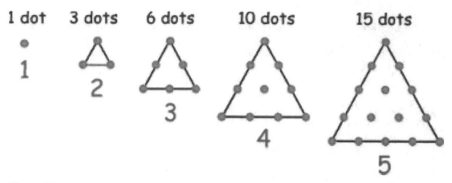

**Figure 24**

The first few triangular numbers are: 1, 3, 6, 10, 15, 21, 28, …. The number 8 plays a special role with regard to these numbers. Multiply any of these numbers by 8 and add 1, the result will always be a square number. For example, if we take the triangular number 15 and multiplied by 8 and add 1, we get $121 = 11^2$. Another example with the triangular number 28, we get: $28 \times 8 + 1 = 225 = 15^2$.

We can often find unusual relationships of numbers such as these consisting of 8s as we can see in Figure 25.

| | | | | |
|---:|:---:|:---|:---:|---:|
| 1 | × | 8 | = | 8 |
| 11 | × | 88 | = | 968 |
| 111 | × | 888 | = | 98568 |
| 1111 | × | 8888 | = | 9874568 |
| 11111 | × | 88888 | = | 987634568 |
| 111111 | × | 888888 | = | 98765234568 |
| 1111111 | × | 8888888 | = | 9876541234568 |
| 11111111 | × | 88888888 | = | 987654301234568 |
| 111111111 | × | 888888888 | = | 9876543190123456 8 |
| 1111111111 | × | 8888888888 | = | 987654321791234568 |

**Figure 25**

In the physical world, we notice that an octagon has eight sides, a cube has eight vertices, and an octahedron has eight faces, as shown in Figure 26. So, the number 8 has some significance in geometry.

**Figure 26**

The number 8 allows us to recognize some unusual characteristics of the number.

- Using the number 8, we can come up with a clever pattern: 8 times the eighth prime $(8 \times 19 = 152)$ results in a number with the sum of the digits is $1 + 5 + 2 = 8$.
- The number comprised of eight ones, $11,111,111 + 8 = 11,111,119$, which is a prime number.
- The difference between 2 prime numbers squared is always a multiple of 8, as for example, $23^2 - 19^2 = 168 = 8 \times 21$.
- We can also generate a prime number by using eight digits of 8: $\dfrac{8^8 + 88888}{8} = 2,108,263$, which is a prime number.
- The number 8 can also generate a prime number as follows: $\dfrac{10^8 - 8}{8} = 12,499,999$, which is a prime number.
- We can also generate a prime number taking an eight-digit number comprised entirely of 8s to the eighth power and adding 1: $88,888,888^8 + 1 = 3,897,443,119,493,995,135,240,117,470,484,161,\\627,805,761,928,038,551,890,318,852,097$, which is a prime number.

A truly amazing curiosity with which to complete our investigation of the number 8 is to recognize that if we take the difference of the sum of the first eight composite numbers and the first eight prime numbers, the difference will be 1:

$2 + 3 + 5 + 7 + 11 + 13 + 17 + 19 = 77$, while $2 + 6 + 8 + 9 + 10 + 12 + 14 + 15 = 76$. Yet another surprise about the number 8.

## THE NUMBER 9

In the American culture the number 9 is used in a number of ways. We say "a cat has 9 lives," or "one is on cloud 9," or "a stitch in time saves 9," or "possession is 9/10 of the law," or "he is dressed to the nines." Someone who wants the entire load says, "I'll take the whole 9 yards." Curiously, the last one refers to 9 cubic yards, which is the entire volume of a cement truck. We should also keep in mind that a baseball team fields nine players at a time for a game that consists of nine innings. Curiously, two famous classical composers—perhaps through superstition—only composed nine symphonies. These are Ludwig von Beethoven and Antonin Dvorak.

The number 9 holds many special properties largely because it is 1 unit less than 10, the base of our number system. Thus, the number 9 is the largest single-digit number. From a "pictorial standpoint," the number 9 can be represented in exponential form as $3^{2^1}$. The number 9 is the only number that is the sum of two consecutive cubes: $9 = 1^3 + 2^3$. The number 9 can also be written as a sum of three consecutive factorials: $9 = 1! + 2! + 3!$. The number 9 also enables a cute pattern: $9^2 = 81$ and $8 + 1 = 9$.

The number 9 also provides us with some amazing patterns such as when we multiply $12,345,679 \times 9 = 111,111,111$. If were to consider multiplying the same number $12,345,679$ by the number $9^2$, we will arrive in another unexpected pattern, namely, $12,345,679 \times 81 = 999,999,999$. Suppose we multiply the reversal of this previous number; however, this time including the number 8, we arrive at another pattern provided by the number 9, as follows $987,654,321 \times 9 = 8,888,888,889$. Analogous patterns can be seen in Figures 27 and 28.

| | | | | | | |
|---:|:---:|:---:|:---:|:---:|:---:|:---|
| **0** | × | **9** | + | **1** | = | **1** |
| **1** | × | **9** | + | **2** | = | **11** |
| **12** | × | **9** | + | **3** | = | **111** |
| **123** | × | **9** | + | **4** | = | **1,111** |
| **1,234** | × | **9** | + | **5** | = | **11,111** |
| **12,345** | × | **9** | + | **6** | = | **111,111** |
| **123,456** | × | **9** | + | **7** | = | **1,111,111** |
| **1,234,567** | × | **9** | + | **8** | = | **11,111,111** |
| **12,345,678** | × | **9** | + | **9** | = | **111,111,111** |

**Figure 27**

The number 9 also provides an opportunity for us to look back at the number 8, which appears repeatedly in Figure 28.

$$0 \times 9 + 8 = 8$$
$$9 \times 9 + 7 = 88$$
$$98 \times 9 + 6 = 888$$
$$987 \times 9 + 5 = 8{,}888$$
$$9{,}876 \times 9 + 4 = 88{,}888$$
$$98{,}765 \times 9 + 3 = 888{,}888$$
$$987{,}654 \times 9 + 2 = 8{,}888{,}888$$
$$9{,}876{,}543 \times 9 + 1 = 88{,}888{,}888$$
$$98{,}765{,}432 \times 9 + 0 = 888{,}888{,}888$$

**Figure 28**

This time will use the 9 in another way to generate an amazing pattern, which is shown in Figure 29.

$$9 \times 9 = 81$$
$$99 \times 99 = 9{,}801$$
$$999 \times 999 = 998{,}001$$
$$9{,}999 \times 9{,}999 = 99{,}980{,}001$$
$$99{,}999 \times 99{,}999 = 9{,}999{,}800{,}001$$
$$999{,}999 \times 999{,}999 = 999{,}998{,}000{,}001$$
$$9{,}999{,}999 \times 9{,}999{,}999 = 99{,}999{,}980{,}000{,}001$$

**Figure 29**

The number 9 also lends itself to some rather playful exercises. Suppose we choose an eight-digit number with no repeating digits and multiply it by 9 to get another number where all nine digits are represented exactly once. We show a few of these examples in Figure 30.

$$81{,}274{,}365 \times 9 = 731{,}469{,}285$$
$$72{,}645{,}831 \times 9 = 653{,}812{,}479$$
$$58{,}132{,}764 \times 9 = 523{,}194{,}876$$
$$76{,}125{,}483 \times 9 = 685{,}129{,}347$$

**Figure 30**

As we know, the decimal expansion of the value of π is infinitely long. However, when we consider the first 1,000 places, as shown in Figure 31, we notice that the number 9 is repeated six times consecutively after the 762nd place. The uniqueness is that no other number in the first 1,000 places was repeated more than three times consecutively. Once again, showing a special appearance of the number 9.

π=3. 1415926535 8979323846 2643383279 5028841971 6939937510 5820974944
5923078164 0628620899 8628034825 3421170679 8214808651 3282306647
0938446095 5058223172 5359408128 4811174502 8410270193 8521105559
6446229489 5493038196 4428810975 6659334461 2847564823 3786783165
2712019091 4564856692 3460348610 4543266482 1339360726 0249141273
7245870066 0631558817 4881520920 9628292540 9171536436 7892590360
0113305305 4882046652 1384146951 9415116094 3305727036 5759591953
0921861173 8193261179 3105118548 0744623799 6274956735 1885752724
8912279381 8301194912 9833673362 4406566430 8602139494 6395224737
1907021798 6094370277 0539217176 2931767523 8467481846 7669405132
0005681271 4526356082 7785771342 7577896091 7363717872 1468440901
2249534301 4654958537 1050792279 6892589235 4201995611 2129021960
8640344181 5981362977 4771309960 5187072113 **4999999**837 2978049951
0597317328 1609631859 5024459455 3469083026 4252230825 3344685035
2619311881 7101000313 7838752886 5875332083 8142061717 7669147303
5982534904 2875546873 1159562863 8823537875 9375195778 1857780532
1712268066 1300192787 6611195909 2164201989

**Figure 31**

To determine if a given number is divisible by 9, one simply needs to find the sum of the digits of this number, and if that digit sum is divisible by 9, then the original number was divisible by 9. For example, suppose we would like to determine if the number 36,180,225 is divisible by 9. We take the sum of the digits of a given number, $3 + 6 + 1 + 8 + 0 + 2 + 2 + 5 = 27$. Although we know that the number 27 is divisible by 9, we could still take this one step further and take the sum $2 + 7 = 9$ and then make the conclusion that the original number was divisible by 9.

We have much to thank Leonardo of Pisa (1170–c. 1245)—more popularly known as Fibonacci—for his brilliance and innovation in a variety of mathematical topics and specifically for introducing the Hindu-Arabic number system to the European world. The numeral system that we use today, namely, 1, 2, 3, 4, 5, 6, 7, 8, 9, and 0 was first introduced to the European world, in Fibonacci's book *Liber abaci*, which was first published in the year 1202. The first sentences in Chapter 1 of this book read as follows: "The nine Indian figures are 9, 8, 7, 6, 5, 4, 3, 2, 1. With these nine figures, and the sign 0, which the Arabs call Zephir, any number whatsoever can be written." It is quite likely that Fibonacci came across these numbers when he spent time in his youth accompanying his father, who was a public official in the Bugia (a port city in eastern Algeria) customs house established for the Pisan merchants. There Fibonacci was taught mathematics using these symbols, which were imported from India, and so he incorporated them into his book, which was republished in 1228. Yet, it is believed that it wasn't until 50 years later that these numerals became more universally used in Europe. We also credit Fibonacci as the first to use the fraction bar to represent a fraction such as $\frac{1}{2}$.

In Chapter 3 of his book *Liber abaci*, Fibonacci introduces a procedure for testing the correctness of an addition computation, which he called "casting out 9's." To check an addition problem, all we need to do is to find the sum of the digits of each of the addends and then take the sum of the addend sums and compare it to the sum of the digits of the computational sum of numbers. If they are the same, then the answer is correct.

We can show this with the following addition computation:

13,579 → digit sum: 25 → digit sum: 7
86,327 → digit sum: 26 → digit sum: 8
**sum of digit sums**: 7 ± 8 = 15 → digit sum: 6
99,906 → digit sum: 33 → digit sum: 6

Because the sum of the digit sums of the two addends eventually is 6, and it matches the digit sum of the addition calculation sum, also 6, the addition could be correct.

The smallest magic square consists of nine cells as shown in Figure 32, where the sum of the numbers in the rows, columns, and diagonals is 15.

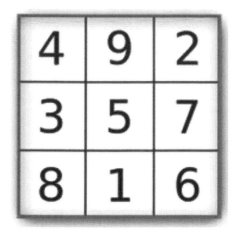

**Figure 32**

This magic square brings with it another curiosity, which is special to this particular nine-cell magic square. If we consider the three-digit numbers formed by the numbers in the rows in both directions, we get the following unusual result:

$$492^2 + 357^2 + 816^2 = 1,035,369 = 294^2 + 753^2 + 618^2.$$

Now consider the columns:

$$438^2 + 951^2 + 276^2 = 1,172,421 = 834^2 + 159^2 + 672^2.$$

Notice how the sums of squares in each case are reversals of one another.

The number 9 also has an interesting place in geometry. Whenever one has three noncollinear points there is always a unique circle that contains the three points. When a fourth point also emerges on the same circle it is quite noteworthy. Imagine that we can show that nine triangle-related points can all lie on the same circle. That is phenomenal! These nine points, for any given triangle, are:

- The midpoints of the sides,
- The feet of the altitudes,
- The midpoints of the segments from the orthocenter[17] to the vertices.

Amazingly all nine points lie on the circle called the *nine-point circle* of the triangle shown in Figure 33.

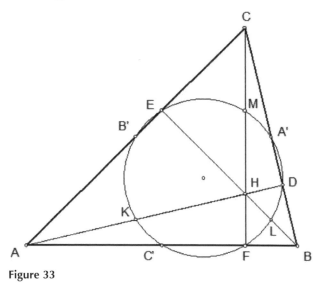

**Figure 33**

In 1765, Leonhard Euler showed that six of these points, the midpoints of the sides and the feet of the altitudes, determine a unique circle. We could have included that amongst our wonders for the number 6, however, we deferred that to when it became even more amazing with nine points on the same circle, which was not discovered until 1820, when a paper[18] published by the French mathematicians Charles-Julian Brianchon (1783–1864) and Jean-Victor Poncelet (1788–1867) appeared, where the remaining three points (the midpoints of the segments from the orthocenter to the vertices) were found to be on this circle. The paper contains the first complete proof of the theorem and uses the name "*the nine-point circle*" for the first time. The theorem states that in any triangle, the midpoints of the sides, the feet of the altitudes, and the midpoints of the segments from the orthocenter (intersection point of the altitudes of the triangle) to the vertices lie on a circle, which is what we today refer to as the nine-point circle. This is a truly remarkable relationship for nine points!

There are many curiosities involving the number 9. For example, the sum of the first nine consecutive prime numbers is a perfect square, as we can see with $2 + 3 + 5 + 7 + 11 + 13 + 17 + 19 + 23 = 100$, which is a perfect square.

The Prussian mathematician Christian Goldbach (1690–1764) conjectured that any odd number greater than 5 can be expressed as a sum of three prime numbers. However, any odd number greater than 9 can be expressed as a sum of three odd prime numbers.

A curiosity involving the number 9 begins with the ninth prime number, 23, when multiplied by 9, yields 207, the sum of whose digits is 9. Also, $10^9 + 9 = 1,000,000,009$, which is a prime number.

The number 9 can also generate prime numbers as follows: $2^9 + 9 = 521$, and $2^9 - 9 = 503$, both of which are primes. Moreover, the number 9 is the smallest composite number that generates primes in this manner. Apparently, the only number that can generate two primes in the following pattern is the number 9 as we can see with: $2^9 + 9^2 = 593$, and $2^9 - 9^2 = 431$, which are both prime numbers.

A view of all prime numbers will expose the fact that the only composite units digits amongst the prime numbers is number 9. Furthermore, when taking the difference of two consecutive squares the number 9 is the only composite units digit that will appear.

The number 9 is the only square number that is equal to the sum of the cubes of two consecutive natural numbers. That is to say, $9 = 1^3 + 2^3$.

We also find that the number 9 it can be expressed in fractional form using all 10 digits exactly once, as we see in the following fractions: $9 = \dfrac{95742}{10638}$, $9 = \dfrac{95823}{10647}$, and $9 = \dfrac{97524}{10836}$. However, if we allow the 0 to take a first position in any of the numbers, we get an additional three fractional equivalents to the number 9, namely, $1,7299 = \dfrac{57429}{06381}$, $9 = \dfrac{58239}{06471}$, and $9 = \dfrac{75249}{08361}$.

Thus, we can see how the number 9 from its place below the number 10 has many unique characteristics.

## THE NUMBER 10

Number 10 holds a very special place in our culture and society. First of all, we have 10 fingers, and the Judeo-Christian belief is based on the Ten Commandments. However, of particular note is that our entire number system is based on the number 10, known as the decimal system. As we mentioned earlier, when we examined the number 6 and noticed it was a triangular number, we now can add to this list of triangular numbers the number 10, which is also a triangular number as we can see in Figure 34. This arrangement should be familiar to anyone who has gone bowling as it is the arrangement of the bowling pins.

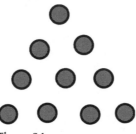

**Figure 34**

When we add the first four numbers, we find that their sum is 10. We see that in our triangular number in Figure 32, where $1 + 2 + 3 + 4 = 10$. We can also reach the number 10 by adding the first three prime numbers: $2 + 3 + 5 = 10$.

The number 10 will also enable us to generate a prime number in the following way:

$$1^1 + 2^2 + 3^3 + 4^4 + 5^5 + 6^6 + 7^7 + 8^8 + 9^9 + 10^{10} = 10,405,071,317,$$

which is a prime number.

Analogously, we can get a prime number with the following series:

$-1! + 2! - 3! + 4! - 5! + 6! - 7! + 8! - 9! + 10! = 3,301,819$, which is the 237,016th prime number.

Another variation of how the first 10 factorials can lead us to a prime number is as follows: $10! + 9! + 8! + 7! + 6! + 5! + 4! + 3! + 2! + 1! + 0! + 1! + 2! + 3! + 4! + 5! + 6! + 7! + 8! + 9! + 10! = 8,075,827$, which is the 544,542nd prime number. A rather unusual aspect of the number $10! = 10 \times 9 \times 8 \times 7 \times 6 \times 5 \times 4 \times 3 \times 2 \times 1 = 3,628,800$, which is exactly the number of seconds in six weeks. While considering factorials, the number 10! provides a rather impressive solution, namely, $10! = 6! \times 7!$, And when wanting to represent the number 10 with three factorials of this sort, we get $10! = 3! \times 5! \times 7!$ Also, $(10!)^2 + 1$ is a prime number, namely, $13,168,189,440,001$. Furthermore, $10! - 11 = 3,628,789$, and $10! + 11 = 3,628,811$, are a pair of twin prime numbers, the 258,689th and the 258,690th primes, respectively.

The number 10 is the only number where the sum and the difference of its prime divisors are both prime numbers because $5 - 2 = 3$, a prime number, and $2 + 5 = 7$, which is also a prime number. Another curious feature of the number 10 is that it is the smallest number whose fourth power can be written as the sum of two squares in two different ways: $10^4 = 80^2 + 60^2$ and $10^4 = 96^2 + 28^2$.

## THE NUMBER 11

In the year 1752, Great Britain and its American colonies removed 11 days in the month of September when converting from the Julian calendar to the Gregorian calendar. Worldwide, the number 11 further plays an important role on the calendar. In 1918, on the 11th day of the 11th month at the 11th hour of the day, there was an armistice ending World War I. We also recall that Apollo 11 was the first spacecraft to land all on the Moon. In American everyday life, we observe that football teams field 11 players of the time, as do soccer, hockey, and cricket teams. For Canadians, the national flag features a maple leaf with 11 points on it. The American $1 coin (is-sued 1979–1981, 1999) is an 11-sided polygon (hendecagon) featuring the portrait of the women's suffrage activist Susan B. Anthony. In the United States the length of a mile is 5,280 feet, or 1,760 yards, or 63,360 inches. Each of these quantities is divisible by 11. We should also recall that it was Apollo 11 mission that put the first man on the moon in 1969, a number which is also divisible by 11.

Arithmetic with the number 11 provides some advantages. *To multiply a two-digit number by 11 just add the two digits and place this sum between the two digits.* Let's try using this technique. Suppose you need to multiply 45 by 11. According to the rule, add 4 and 5 and place it between the 4 and 5 to get 495. This can get a bit more difficult when the sum of the two digits you added results in the two–digit number. What do we do in a case like that? We no longer have a single digit to place between the two original digits. If the sum of the two digits is greater than 9, then we place the units digit between the two digits of the number being multiplied by 11 and "carry" the tens digit to be added to the hundreds digit of the multiplicand.[19] Let's try it with $78 \times 11$. As before, we add the two digits $7 + 8 = 15$. We place the 5 between the 7 and 8, and add the 1 to the 7, to get $[7 + 1][5][8]$ or 858.

The technique also holds when 11 is multiplied by a number of more than two digits.

For example, for a larger number such as 12,345 we can multiply it by 11 as before, we begin at the right-side digit and add every pair of digits going to the left. $1[1 + 2][2 + 3][3 + 4][4 + 5]5 = 135,795$.

If the sum of two digits is greater than 9, then use the procedure described in the preceding text: place the units digit appropriately and carry the tens digit. As an example, multiply 456,789 by 11.

We carry the process step by step as shown in Equation 3.

$$4[4+5][5+6][6+7][7+8][8+9]9$$

$$4[4+5][5+6][6+7][7+8][17]9$$

$$4[4+5][5+6][6+7][7+8+1][7]9$$

$$4[4+5][5+6][6+7][16][7]9$$

$$4[4+5][5+6][6+7+1][6][7]9$$

$$4[4+5][5+6][14][6][7]9$$

$$4[4+5][5+6+1][4][6][7]9$$

$$4[4+5][12][4][6][7]9$$

$$4[4+5+1][2][4][6][7]9$$

$$4[10][2][4][6][7]9$$

$$[4+1][0][2][4][6][7]9$$

$$[5][0][2][4][6][7]9$$

$$5,024,679$$

**Equation 3**

The number 11 lends itself to a very clever technique for determining whether a given number is divisible by 11. The technique is as follows: *take the sum of the odd-placed digits of the original number. Then take the sum of the remaining digits. If the difference of these two sums is a multiple of 11, then the original number was a multiple of 11.* For example, consider the number 1,357,531. The sum of the odd-placed digits is $1 + 5 + 5 + 1 = 12$. The sum of the remaining digits is $3 + 7 + 3 = 13$. The difference of these 2 sums is $13 - 12 = 1$, which is not a multiple of 11. Therefore, the original number, 1,357,53, is not a multiple of 11. However, consider number 135,795 and take the sum of the odd placed digits $1 + 5 + 9 = 15$, and the sum of the remaining digits $3 + 7 + 5 = 15$. The difference of the sums is $15 - 15 = 0$, which is a multiple of 11 and, therefore, the original number is a multiple of 11.[20]

There is another way to determine if a number is divisible by 11. Take the given number that you are testing for divisibility by 11 and break it up into pairs of adjacent digits. If number consists of an odd-number of digits, place a zero on either end of the number. Find the sum of these pairs of two-digit numbers. If the resulting number is divisible by 11, then the original number was divisible by 11. Let's consider the following example: suppose we would like to test to see if the number 786,432,614 is divisible by 11. We partition this long number into pairs of consecutive digits, as follows: $07 + 86 + 43 + 26 + 14 = 176$ continuing this process: $01 + 76 = 77$, which is a multiple of 11, and therefore the original number was divisible by 11.

There are many curiosities embedded in the number 11, such as taking any number that is a multiple of 11 and reversing the digits, the result will be another number that is also a multiple of 11. For example, the number $135,795 = 11 \times 12,345$. Reversing the digits of the original number to get $597,531 = 54,321 \times 11$. Notice that the number and the multiplier are reversals of the original product.

The number 11 lends itself to some very curious mathematical surprises. Suppose you have a number of multiple digits where no two adjacent digits have a sum greater than 9 then you can do the following:

- Multiply the number by 11,
- Reverse the digits of this product,
- Divide the resulting number by 11,
- And you will find that you have attained a number whose digits are the reverse of the original number.

For example, consider a sample number and multiply it by 11 to get $36,181,451 \times 11 = 397,995,961$. We now reverse the digits and divide that number by 11 to get: $169,599,793 \div 11 = 15,418,163$, which unexpectedly is a reversal of the original number.

Aside from being the fifth prime number, it is also the fifth Lucas number. You may recall the Lucas numbers form a sequence of numbers beginning with 1 and 3 where each succeeding number is the sum of the two previous numbers, as in the following sequence: 1, 3, 4, 7, 11, 18, 29, 47, 76, 123, .... The sequence was popularized by the French mathematician Edouard Lucas (1842–1891), who also brought popularity to the Fibonacci numbers from which he got the idea of this sequence.[21]

We also find the powers of 11 on the first few rows of the famous Pascal triangle as shown in Figure 35.

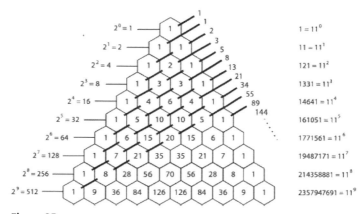

**Figure 35**

Up to the fifth row the powers of 11 appear directly. To get the fifth power of 11 we notice that this row has two-digit numbers, so we need to carry over the tens digit of each of these two-digit numbers to the next place to get $11.^5$

To the left of the Pascal triangle, you will notice that the sums of the numbers generate the powers of 2, while the oblique sums show the Fibonacci numbers, which are similar to the Lucas numbers in that they are also generated by the sum of consecutive numbers; however, this time beginning with 1 and 1 as the first two numbers. The Fibonacci numbers also present an appearance of the number 11 in a rather camouflaged fashion. Again, somewhat unexpectedly, the sum of any 10 consecutive Fibonacci numbers is divisible by 11. For example, suppose we take the following 10 consecutive Fibonacci numbers: $5 + 8 + 13 + 21 + 34 + 55 + 89 + 144 + 233 + 377 = 979 = 11 \times 89$.

When the number 11 is placed between other multiples of 11, the results are multiples of 11, as we can see with the following numbers: 1,111, 2,112, 3,113, 4,114, 5,115, 6,116, 7,117, 8,118, and 9,119.

- Here is a cute way of generating an 11-digit number which begins and ends with a number 11: we begin with a repunits number with 11 digits and then subtract 11-factorial: $11,111,111,111 - 11! = 11,071,194,311$, which is a prime number. Once again, we find that the number, 11, that is one greater than the number of the base, has many unique properties for us to consider.
- First of all, the number 11 is the only even-digit prime number that is a palindrome.
- The number 11 is also the only prime number that can be expressed in the following way using two consecutive primes: $2^3 + 3 = 8 + 3 = 11$ and $3^2 + 2 = 9 + 2 = 11$.
- We can create an 11-digit prime number by taking the first 11 numbers as follows: $12,345,678,910 - 11 = 12,345,678,899$, which is a prime number.
- The number 11 can also be used to generate prime numbers in the following ways:
$$\frac{11^{11} \times 2^{11} + 1}{2^{11} - 1} = \frac{584318301411329}{2047} = 285451051007. \text{ This is a prime}$$
number.
- Furthermore, using all the prime numbers up to the number 11 in the following way generates a prime number: $2 + 3^5 \times 7^{11} = 480,490, 398,551$, which is a prime number.
- Also, the number consisting of 11 digits: $11,111,111,111 = 21,649 \times 513,239$, generates a product of two prime numbers.

- We can also use 11 ones (disguised as other numbers) as follows to generate a prime number: $11^{11} + (11^{11} \times 11) - 1 = 3,423,740,047,331$, which is a prime number. Another way in which 11 ones can generate a prime number is as follows: $11! + 11! + 11! + 11! + 11! + 1 = 199,584,001$, which is a prime number.
- Two consecutive prime numbers can be generated as follows: $10! - 11 = 3,628,789$, and $10! + 11 = 3,628,811$.
- The number 11 is the only prime number where if you add or subtract 6 or 8 from 11 you will always end up with a prime number, as we can see with the following:
  $11 + 6 = 17, 11 - 6 = 5, 11 + 8 = 19$, and $11 - 8 = 3$.
- The number 11 is the smallest prime number that can be expressed as the sum and product of two other prime numbers such as: $(2 + 3) + (2 \times 3) = 11$.
- The number 11 squared can be expressed as a sum of three consecutive prime numbers: $11^2 = 121 = 37 + 41 + 43$.
- The number 11 cubed can be expressed as a sum of three consecutive prime numbers as well:

  $11^3 = 1,331 = 439 + 443 + 449$.

- The sum of the next 11 primes after the number 11, namely, $13 + 17 + 19 + 23 + 29 + 31 + 37 + 41 + 43 + 47 + 53 = 353$, which is a palindromic number, and the 71st prime number.
- We can also express the number 11 as the sum of a square and a prime in two different ways—and, by the way, it is the smallest number that has this property: $11 = 2^2 + 7$ and $11 = 3^2 + 2$.

The curiosities with the number 11 are offered to amuse the reader further about this number amazing number. First, we have $11^2$ it as the sum of five consecutive powers of 3 as follows:
  $11^2 = 121 = 3^0 + 3^1 + 3^2 + 3^3 + 3^4$. Then we have the $11^3$ as the sum of the squares of three consecutive odd numbers: $11^3 = 1331 = 19^2 + 21^2 + 23^2$. There are many more for the motivated reader to pursue.

## THE NUMBER 12

For most people, the number 12 indicates the number of months in a year, or the number of hours on the face of an analog clock. The number 12 is quantified by the word *dozen*. In the United States, 12 inches is equal to 1 foot. The number 12 also appears in religious history. For example, in ancient days there were 12 tribes of Israel. In the Christian religion, there were 12 Apostles and there are "12 days of Christmas." The number 12 also permeates the arts, such as in theater. "The 12th Night" is a famous comedy by William Shakespeare (1564–1616). In music, the 12-tone music was developed by the Austrian composer Arnold Schoenberg (1874–1951). Yet, many may also recall that the earliest phonograph records were referred to as 12-inch records because that described their diameter.

In mathematics, the number 12 has some peculiarities. For example, 12 is divisible by the sum of its digits and by the product of its digits. The number 12 is also the largest known even number that can be expressed as a sum of two prime numbers in only one way because $5 + 7 = 12$.

Perhaps even more impressive is that the sum of any pair of twin primes (prime numbers differ by 2) is divisible by 12. For example, the twin prime pair, 29 and 31, whose sum is 60 is divisible by 12.

Here is a true coincidence, where the 12th prime, 37, is a reversal of the 21st prime, 73. Notice both reversals!

Furthermore, the number 12 is a factor of 1 less than any prime number (greater than 3) squared. Here are a few examples: $11^2 - 1 = 120 = 10 \times 12$, or $23^2 - 1 = 528 = 44 \times 12$, or $37^2 - 1 = 1,368 = 114 \times 12$. You may wish to try others to convince yourself that this is true for all primes greater than 3.

The number 12 also has a peculiarity that is shared by very few other numbers. Consider $12^2 = 144$, and the square of the number with its digits reversed $21^2 = 441$. The results are reversals of one another. We shall revisit this pattern with the number 13.

Generating prime numbers in unusual ways with the number 12:

- We know that $2 \times 2 \times 3 = 12$, but now following this pattern, using 12 as an exponent, we get: $2^{12} + 2^{12} + 3^{12} = 539,633$, which is a prime number.
- Consider the five-digit number consisting of five consecutive digits 34,567, when added to 12! a prime number is arrived at, namely, 479,036,167.

In geometry, the number 12 also has a special place among the five Platonic solids, where the dodecahedron has 12 congruent pentagonal faces, the icosahedron has 12 vertices, the octahedron and the cube each have 12 edges, as we can see in Figure 36.

| Platonic Solids | | Vertices | Edges | Faces |
|---|---|---|---|---|
| tetrahedron | | 4 | 6 | 4 |
| cube | | 8 | 12 | 6 |
| octahedron | | 6 | 12 | 8 |
| dodecahedron | | 20 | 30 | 12 |
| icosahedron | | 12 | 30 | 20 |

**Figure 36**

## THE NUMBER 13

The number 13 is usually associated with being an unlucky number. Buildings with more than 13 stories will typically omit the number 13 from the floor numbering. This is immediately noticeable in an elevator, where there is no button for 13. You can certainly think of other examples where the number 13 is associated with bad luck such as when the 13th of a month turns up on a Friday, then it is considered a particularly a bad omen. This may derive from the belief that there were **13** people present at the Last Supper, which resulted in the Jesus' crucifixion on a **Friday**. Do you think that the 13th comes up on a Friday with equal regularity as on the other days of the week? You may be astonished that, lo and behold, the 13th comes up a bit more frequently on Friday than on any other day of the week.

This fact about Friday the 13th was first published by B. H. Brown,[22] where he stated that the Gregorian calendar follows a pattern of leap years, repeating every 400 years. The number of days in one four-year cycle is $3 \times 365 + 366$. Therefore, in 400 years there are

$100 \times (3 \times 365 + 366) - 3 = 146,097$ days. (Note that century years that are divisible by 400 are not leap years, hence the deduction of 3.) This total number of days is exactly divisible by 7. Because there are 4,800 months in this 400-year cycle, the 13th comes up 4,800 times according to the following table. Interestingly enough, the 13th comes up on a Friday more often than on any other day of the week. The following chart summarizes the frequency of the 13th appearing on the days of the week.

| Day of the Week | Number of 13s | Percent |
| --- | --- | --- |
| Sunday | 687 | 14.313 |
| Monday | 685 | 14.271 |
| Tuesday | 685 | 14.271 |
| Wednesday | 687 | 14.313 |
| Thursday | 684 | 14.250 |
| *Friday* | *688* | *14.333* |
| Saturday | 684 | 14.250 |

Then there are well-known triskaidekaphobics[23], such as Napoleon Bonaparte, Herbert Hoover, Mark Twain, Richard Wagner, and Franklin Delano Roosevelt. It was said that Roosevelt would not invite 13 people around the dinner table and would not begin any trip on the 13th of the month. Richard Wagner (1813–1883), the famous German composer who revolutionized music, might have attributed his phobia for the number 13 to the Norse legends where 12 gods were gathered in Valhalla and when the 13th troublesome god, Loge, arrived and caused deadly mischief, the number 13 was seen as unlucky. To what extent Wagner was influenced by this legend will remain unknown; however, for him the number 13 seemed to have played a curious role. Wagner included Loge in his famous *Der Ring des Nibelungen*, although not as a bad person, but merely as a manipulator. To understand how the number 13 was manifested in Wagner's life consider the following:

Wagner was born in 1**813**, which has a digit sum of **13**.

Wagner died on February **13**, 1883, which was the **13**th year of the unification of Germany.

Wagner wrote **13** operas during his lifetime.

Richard Wagner's name consists of **13** letters.

Wagner's opera, Tannhäuser, was completed on April **13**, 1845.

On May **13**, 1861 Wagner premiered Tannhäuser in Paris during his **13**-year exile from Germany.

The grand opening of Wagner's festival opera house in Bayreuth, Germany was opened on August **13**, 1876.

Wagner completed his last opera, Parsifal, on January **13**, 1882.

Wagner's last day in Bayreuth, the city in which he built his famous festival opera house, was September **13**, 1882.

On a more positive note, Ludwig van Beethoven's Pathetique Sonata is Opus No. 13. We also know that there are 13 cards in each suit of a deck of playing cards. At age 13, in the Jewish religion, a boy becomes a man at a bar mitzvah, clearly a positive acceptance of the number 13.

The original flag of the United States of America had 13 stripes and 13 stars to commemorate the initial 13 colonies that formed the United States. Today's flag has retained the 13 stripes but now has the number of stars representing the 50 states.

However, on the brighter side, in mathematics the number 13 is sometimes referred to as a "star number" because 13 dots can be arranged in the shape of a hexagram or six-corner star, as we can see in Figure 37.

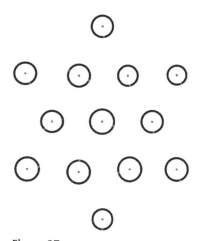

**Figure 37**

In mathematics there are also some curious specialties such as the pattern we see here: $13^2 = 169$ and $31^2 = 961$. It is also curious that the prime number 13 is one of those numbers that when it is reversed is also a prime number, namely, 31. Furthermore, the number 13 is the smallest number that can be expressed as the sum of two prime numbers squared, namely, $2^2 + 3^2 = 13$.

The number 13 is one of three Wilson primes, where the other two such primes are the numbers 5 and 563. A Wilson prime, named for the British mathematician John Wilson (1741–1793), is a prime number $n$, which is a factor of $(n - 1)! + 1$. In this case, when $n = 13$, we have $(13 - 1)! + 1 = 479,001,600 + 1 = 479,001,601 = 13 \times 36,846,277$. We can even take this a step further by noticing that $(13 - 1)! + 1$ also has $13^2$ as a factor. Namely $479,001,601 = 13^2 \times 2,834,329$. While we are considering $13^2$, we can notice that it is the difference of two consecutive cubes as follows: $13^2 = 169 = 8^3 - 7^3$.

Despite its negativity in cultural contexts, the number 13 still has a few redeeming values from a mathematical viewpoint.

Powers related to 13 can lead to some interesting relationships. Some such are as follows:

- $13^{13} - 13 + 1 = 3,028,751,046,592,241$ is a prime number.
- $13^2$ is a factor of $12! + 1 = 479,001,601 = 13^2 \times 2,834,329$.
- $13^2 = 7 + 8 + 9 + 10 + 11 + 12 + 13 + 14 + 15 + 16 + 17 + 18 + 19$.
- $13^2 = + 8^3 + 7^3$.
- $2^{13} - 13 = 8179$, which is a prime number.
- $13 \times 2 = -1^3 + 3^3 = 27 - 1 = 26$.
- We can use the first three prime numbers to represent 13 as: $2^3 + 5 = 13$.
- The sum of the first six prime numbers: $2 + 3 + 5 + 7 + 11 + 13 = 41$, is the 13th prime number.
- If we divide 13 by each of the first 13 prime numbers, the sum of the remainders $1 + 1 + 3 + 6 + 2 = 13$.
- This number 13 is the smallest prime number where the sum of his digits is a perfect square, namely, 4.
- The sum of the first 13 prime numbers $2 + 3 + 5 + 7 + 11 + 13 + 17 + 19 + 23 + 29 + 31 + 37 + 41 = 238$, whose digit sum is 13.
- A prime number can be generated by applying the number 13 in a rather curious way. The numerator of the following fraction has 14 digits of which 13 digits are 3s: $\dfrac{13333333333333}{13} = 1,025,641,025,641$, which is a prime number.
- There are exactly 13 prime numbers consisting of four consecutive digits, and they are: 1,423, 2,143, 2,341, 2,543, 4,231, 4,253, 4,523, 4,567, 4,657, 5,647, 5,867, 6,547, 6,857.
- The number of composite numbers between 13 and its reversal, 31, is 13. These composite numbers are: 4, 6, 8, 9, 10, 12, 14, 15, 16, 18, 20, 21, 22, 24, 25, 26, 27, 28.
- We can represent the number 13 in terms of just prime numbers as follows: $(5 \times 11) - (2 \times 3 \times 7)$. A motivated reader might like to try to create the number 13 in other ways using only prime numbers.
- The number 13 is the only prime number where its double, 26, increased by 1 is a perfect cube: $(2 \times 13) + 1 = 27 = 3^3$.
- A nifty palindrome can be created which begins and ends with 13.131, 211,109,876,543,212,345,678,910,111,213. What makes this palindrome special is that it is a prime number.

In geometry there is a unique right triangle, with hypotenuse length 13, where the area is numerically equal to the perimeter. The sides of this triangle are 5, 12, and **13**. The area of this triangle is $\frac{1}{2} \times 5 \times 12 = 30$, while the perimeter is equal to $5 + 12 + 13 = 30$.

The Pythagorean theorem states that $13^2 = 5^2 + 12^2$. This is the only such triple, which can be transformed into another Pythagorean triple by placing the digit 1 to the left of each of these three numbers to get: $113^2 = 15^2 + 112^2$.

## NUMBER 14

In the United Kingdom, the number of days that comprise a "fortnight" is 14. In the British weight units, the number of pounds comprising a "stone" is 14. Furthermore, Ludwig van Beethoven's famous *Moonlight Sonata* is Piano Sonata No. 14. The 14th state to join the United States was the state of Vermont on March 1, 1791. There isn't too much more that can be said about the number 14, yet it does serve a role in some high-rise buildings, where "14" replaces the 13th floor designation in buildings where the superstition of 13 being an unlucky number dominates.

When we find the sum and the difference of the digits of the number 14, we come up with a pair of twin primes, namely, 3 and 5.

A number curiosity surrounding the number 14 is that the 14th prime number (43) can be used to generate a prime number along with the number 14, as we see with the following:

$14^{14} + 43$ is a prime number, namely, 11,112,006,825,558,059.

The number 14 is the smallest number whose prime factors, 2 and 7 have a sum that is a perfect square, namely, 9.

In mathematics, it partners with the numbers 13 and 15 forming a triangle whose sides are 13, 14, 15, which has a very special property, namely, that the numerical value of its area is twice the numerical perimeter. Using Hero's formula $\left( Area = \sqrt{s(s-a)(s-b)(s-c)} \right)$, where $s$ is the semiperimeter of the triangle, whose sides are lengths $a$, $b$, $c$, to find the area of this triangle we get: $Area = \sqrt{21(8)(7)(6)} = \sqrt{7056} = 84$. The perimeter is $13 + 14 + 15 = 42$.

This provides a good segue to our next number, 15.

## THE NUMBER 15

In the British game of rugby each team has 15 players on the field at any given moment during the game. In the American game of football each quarter of the game has 15 minutes duration. The 15th state to join United States was the state of Kentucky on June 1, 1792.

The number 15 tends to outshine its predecessor, in that it is a triangular number and a hexagonal number as shown in Figure 38. Another way of looking at that is that the sum of the first five natural numbers is equal to 15, as $1 + 2 + 3 + 4 + 5 = 15$.

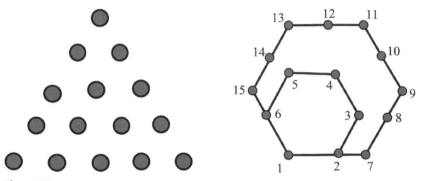

**Figure 38**

The number 15 is also the product of the first two prime numbers, namely, $3 \times 5 = 15$. Because the numbers 3 and 5 are each Fermat primes, which means they can be expressed as $F_n = 2^{2^n} + 1$, (i.e., $3 = 2^{2^0} + 1 = 2 + 1$, and $5 = 2^{2^1} + 1 = 4 + 1$) a regular polyhedron of 15 sides is then constructable using only an unmarked straight-edge and a pair of compasses.

Furthermore, when partnering with the next larger triangular number, 21, we find that 15 and 21 form the smallest pair of triangular numbers, where sum and difference, 36 and 6, are also triangular numbers. We have to go rather far along to find the another such pair of triangular numbers, namely, 780 and 990, whose sum, 1,770 and difference, 210, both of which are also triangular numbers.

Recall our earlier focus on the number 3, where we described the $3 \times 3$ magic square. The sum of the numbers in each of the rows and columns and diagonals is 15, as you can see in Figure 39.

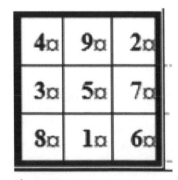

**Figure 39**

The number 15 also plays an important role in a strategy puzzle referred to as the "15 puzzle." This puzzle, pictured in Figure 40, bears its name from the number of square tiles that need to be moved around to obtain numerical order.

**Figure 40**

The number 15 also provides for some unusual patterns such as $2^{15} - 51$ is a prime number.

Also, the number 15 is the smallest number that can be used in the following procedure to generate primes: $(15 \times 4) + 1 = 61$, which is a prime number, and $(15 \times 4) - 1 = 59$, is also prime number.

Starting with a number 15, we can generate a surprising prime number as follows:

$$15! - 14! + 13! - 12! + 11! - 10! + 91 - 8! + 7! - 6! + 5! - 4! + 3! - 2! + 1!$$
$$= 1,226,280,710,981,$$ which is truly a prime number.

A motivated reader we wish to verify that there are exactly 15 palindromic primes each comprised of three digits.

## THE NUMBER 16

In the United States, when girls reached the age of 16, they typically celebrate their "sweet 16" birthday party as a signal of transition into womanhood. Several countries issue 16-year-olds with adult passports, which then typically last 10 years. At the age of 16 in various countries students can join the military, are permitted to vote, can donate blood, and so forth. There are exactly 16 ounces in 1 pound. The number 16 played a role in the film industry where the films were 16mm in size. Moreover, many credit cards numbers are 16 digits long. So, we can see how the number 16 seems to be present in our everyday lives. It should be noted that the state of Tennessee was the 16th state to join the United States on June 1, 1796.

In geometry, the number 16 also plays a unique role when we consider a square with side length 4, its perimeter and area are numerically equal to the number 16. While considering a square with 16 equal cells, as shown in Figure 41, one can easily create the magic square that was made famous by the German artist Albrecht Durer (1471–1528), in his well-known etching entitled "Melancholia I," which is shown in Figure 42.

| 16 | 3 | 2 | 13 |
|----|----|----|----|
| 5 | 10 | 11 | 8 |
| 9 | 6 | 7 | 12 |
| 4 | 15 | 14 | 1 |

**Figure 41**

**Figure 42**

The magic square provides an arrangement of the 16 numbers, where each row, column, and diagonal have a sum of 34. However, there is a wealth of amazing relationships to be seen in this arrangement of 16 numbers, some of which can be seen in the book *Numbers: Their Tales, Types, and Treasures* (A. S. Posamentier and B. Thaller, Prometheus Books, 2015).

The number 16 also has some mathematical curiosities. It is well known that the number 16 is a square number, $4^2 = 16$, but it is also the second fourth-power number after the number 1, that is, $2^4 = 16$.

Also, it is the smallest number that can be expressed in two different ways as a sum of two prime numbers, as $16 = 5 + 11 = 3 + 13$. Additionally, the number 16 is also the smallest square that can be expressed as the difference of two prime squares, namely, $5^2 - 3^2 = 16$. Another curiosity about the number 16 is that it is the smallest square number where the sum of its digits is a prime number, 7.

The number 16 is also the smallest square number whose reversal is a prime number, namely 61.

If we want to be "artistic" to represent the number 16 as a generator of a prime number, we could write it as follows: $16^{1+6} + (1 + 6) = 268,435,463$, which is a prime number.

There are times when we search in unusual ways for the presence of a number as is the case here with the number 16: The sum of the 16th prime number (53) and the 16th composite number (26) is a number, $53 + 26 = 79$, the sum of its digits is 16, as $7 + 9 = 16$.

We also can find that the number 16 is the first square number can be expressed as the sum of two triangular numbers in two different ways, such as $16 = 6 + 10 = 1 + 15$, where 1, 6, 10, and 15 are all triangular numbers. The number 16 also lends itself to a rare situation in which it is the only square integer that can be expressed as $x^y = y^x$, in that $2^4 = 4^2$. This uniqueness was proved by the famous Swiss mathematician Leonhard Euler (1707–1783).

There is an interesting pattern for generating prime numbers, where the number 16 is the largest number that allows this to happen. We first add 16 to itself and subtract 1, then add half of 16 to itself and subtract 1, and so on. Each of these will result in a prime number as follows:

$16 + 16 - 1 = \mathbf{31}$, $16 + 8 - 1 = \mathbf{23}$, $16 + 4 - 1 = \mathbf{19}$, $16 + 2 - 1 = \mathbf{17}$,
    each of these is a prime number.

As a tribute to its uniqueness, the number 16 is the largest known integer, $n$, where the number $2^n + 1$ is a prime number. In other words, $2^{16} + 1 = 65,536 + 1 = 65,537$ is the 6,543rd prime number, and no greater value of $n$ would produce a larger prime number.

## THE NUMBER 17

When it comes to the number 17, one of the author's favorite films is the 1953 classic film *Stalag 17*, for which William Holden won an Oscar award as best actor. Yet, the number 17 finds itself in the title of many songs and other films such as the 1932 film *Number 17* directed by Alfred Hitchcock. We further encounter this curious number 17 when the Armed Forces of the United States offers a "17-gun salute" to honor high-level officers. Yet, in the Italian culture, for variety of reasons, the number 17 has gotten to be an unlucky number. Each side of the Parthenon in Athens, Greece has 17 columns from front to back. It was the 17th parallel that divided North and South Vietnam (as defined by the 1954 Geneva Conference) and lasted until July 1976 when the two parts of Vietnam were united. The state of Ohio was the 17th state to join the United States on March 1, 1803. A 17-jewel watch is one when jewels (usually man-made rubies) are placed at pivot points to make the watch run smoothly.

In mathematics, the number 17 possesses some unusual properties. Not only is the number 17 a prime number, but it is also the sum of the first four prime numbers, namely, $2 + 3 + 5 + 7 = 17$. It should be noted that this is the only time that the sum of four consecutive prime numbers could result in another prime because any of the other four consecutive primes would not include an even number, and therefore, would result in a sum of four odd numbers, which is an even number and, therefore, not a prime number, as the only even prime number is the number 2.

The number 17 is a smallest prime number the sum of whose digits is a perfect cube because $1 + 7 = 8 = 2^3$. The number 17 can also be expressed as the sum of two fourth powers, making it the smallest prime with this feature. That is, $1^4 + 2^4 = 1 + 16 = 17$. Moreover, the number 17 is also the smallest prime number that can be expressed as the sum of the squared prime and a cubed prime, namely, $3^2 + 2^3 = 9 + 8 = 17$. This is the only prime number where such consecutive primes can be used in this way.

The number 17 is a smallest prime, which is the average of seven consecutive prime numbers, namely, 7, 11, 13, 17, 19, 23, 29. We get the average as: $\dfrac{7 + 11 + 13 + 17 + 19 + 23 + 20}{7} = \dfrac{119}{7} = 17$. We should also note that the number 17 is the smallest prime that is the average of two squared primes, namely, $\dfrac{3^2 + 5^2}{2} = \dfrac{9 + 25}{2} = \dfrac{34}{2} = 17$.

The number 17 is also the first member of a twin prime with the number 19, as they are consecutive primes. If you place a 9 between each of the digits of these two numbers, we get 197 and 199, which is another pair of twin primes.

Although $2^{17} = 131,072$ is not a prime number, its reversal, 270,131, is indeed the 23,652nd prime number.

Here is a curiosity that is quite astonishing: $17^7 + 17^4 + 17^1 + 17^3 + 17^2 + 17^4 + 17^5 + 17^6 + 17^5 + 17^8 = 7,413,245,658$. Notice the connection of the exponents with the digits of the sum!

There are times when we can creatively highlight the number 17. Consider the following: $2^{17+17} = \mathbf{17,179},869,184$, notice the first four digits!

A repunit number with 16 ones, 1,111,111,111,111,111 is divisible by 17, in other words, $\dfrac{1,111,111,111,111,111}{17} = 65,359,477,124,183$.

The Pythagoreans were enchanted with the two numbers 16 and 18 and despised the number 17 that separated them. However, the number 17 has some properties that give it some importance as well. The number 17 is the seventh prime number, and 17 generates the sixth Mersenne number[24] 131,071. Furthermore, 17 is the sum of the first four prime numbers: $2 + 3 + 5 + 7 = 17$.

Among prime numbers number 17 also plays a rather unusual role. The expression $P = n^2 + n + 17$, when taking on the values of $n$ from 0 to 15, will always yield a prime number, the smallest of which is when $n = 0$, it yields the number 17. The largest prime yielded for $P$ is when $n = 15$, then $P = 257$. The ambitious reader may wish to try finding the value of $P$ for larger numbers such as $n = 16$, where $P = 289$, which is obviously not a prime number because $289 = 17^2$.

Recall the famous Fibonacci numbers: $1, 1, 2, 3, 5, 8, 13, 21, 34, 55, 89, \ldots$. The number 17 is the only prime number that is the average of two consecutive Fibonacci numbers, namely 13 and 21, because $\dfrac{13 + 21}{2} = 17$. By the way, there are no odd Fibonacci numbers that are divisible by 17. Furthermore, the number 17 is a smallest prime number, which is *not* the sum of consecutive Fibonacci numbers. We can also note that 17 is the only prime number that is half of a Fibonacci number, namely, 34.

The number 17 can also generate 11 prime numbers with values of $n$ from 0 to 10 in the following polynomial: $n^3 + n^2 + 17$.

Furthermore, the number 17 is also considered a Fermat prime number as it satisfies the requirement, $2^{2^n} + 1$, when $n = 2$ because $2^{2^2} + 1 = 17$. This will allow for the construction of a 17-sided polygon, a heptadecagon, with an unmarked straightedge and a pair of compasses. This was one of the proudest findings of the German mathematician Carl Friedrich Gauss (1777–1855), who at age 18 then claimed that this finding motivated him to pursue the study of mathematics and further requested that it be engraved on his gravestone.

The number 17 is the smallest number that can be written as the sum of a squared number and a cubed number in two different ways: $1^3 + 4^2 = 1 + 16 = 17$, and $2^3 + 3^2 = 8 + 9 = 17$. The number 17 is also the smallest number that can be expressed as a sum of two fourth powers, namely, $1^4 + 2^4 = 17$.

The number 17 also has some strange characteristics. For example, $17^3 = 4,913$ and the sum of the digits of this number is: $4 + 9 + 1 + 3 = 17$. By the way, the only other numbers that share this characteristic are $1, 8, 18, 26$ and $27$. You might want check this to be convinced of this property. We offer one here: $26^3 = 17,576$, and $1 + 7 + 5 + 7 + 6 = 26$.

Some prime numbers, when their digits reversed, also deliver prime numbers. As you can see from the list of the first few of these, the number 17 is one of these unusual numbers:

13, 17, 31, 37, 71, 73, 79, 97, 107, 113, 149, 157, …

Here is a challenge for an interested reader: letting the letters of the alphabet take their numerical position, such as A = 1, B = 2, C = 3, …Y = 25, Z = 26, show that the word SEVENTEEN is a smallest number whose value is a prime number because $19 + 5 + 22 + 5 + 14 + 20 + 5 + 5 + 14 = 109$, a prime number.

As a further challenge to the motivated reader, show how the number $17^2$ can be expressed as the sum of $1, 2, 3, 4, 5, 6, 7, 8$ distinct square numbers.


## THE NUMBER 18


A curiosity can be experienced with the two 18-letter words: "conservationalists" and "conversationalists," which are anagrams of each other. They are considered the longest pair of anagrams in the English language—if scientific words are excluded.[25]

In the Chinese culture number 18 represents a word that means that someone will prosper. In the Jewish culture, the number 18 is classically seen as a good luck number. The reason for this is that when using Hebrew letters to represent the number 18. It appears as חי (pronounced chai), which as a word spells out "life," and is often seen worn by women as a locket. In sports, the number 18 manifests itself in defining a full golf course. The 18th state to join the United States was the state of Louisiana on April 30, 1812.

We can also have some fun with the number 18. Begin by taking any three-digit number whose digits are all different, and arrange them to form the largest number and then the smallest number. Subtract the smaller number from the larger number, and you will find that this answer will be a number, the sum of whose digits is 18. Let's consider an example. We shall select the number 584. We then write the smallest number with these digits (458) and then the largest number with these digits (854). Subtracting these numbers 854 − 458, we get 396. The sum of the digits of this number $(3 + 9 + 6)$ is 18. Try it with some other three-digit number to convince yourself of this property.

The number 18 does not harbor too many mathematical curiosities. However, one such is that $18 = 9 + 9$, and $9 \times 9$ is equal to its reversal, 81. As we search for more peculiarities, we can notice that 18 is the smallest number that is twice the sum of its digits. One can generate two operations with the number 18 that will yield two numbers that use all 10 digits exactly once as in the following: $18^3 = \mathbf{5,832}$, and $18^4 = \mathbf{104,976}$. When we consider the third power, the sixth power, and the seventh power of 18, we find that each provides a number whose digits have a sum of 18, as shown in the following text:

$18^3 = 5,832$ and $5 + 8 + 3 + 2 = 18$,

$18^6 = 34,012,224$, and $3 + 4 + 0 + 1 + 2 + 2 + 2 + 4 = 18$

$18^7 = 612,220,032$, and $6 + 1 + 2 + 2 + 2 + 0 + 0 + 3 + 2 = 18$

Just to take this to a "higher" level, we find that

$18^{18} = 39,346,408,075,296,537,575,424$, and $3 + 9 + 3 + 4 + 6 + 4 + 0 + 8 + 0 + 7 + 5 + 2 + 9 + 6 + 5 + 3 + 7 + 5 + 7 + 5 + 4 + 2 + 4 = 108$.

The number 18 is one of two numbers where a rectangle has a numerical measure of its area equal to its perimeter. This rectangle is one where the length and width are 6 and 3, respectively, and where the area is $3 \times 6$ and the perimeter is $3 + 3 + 6 + 6$, each of which is equal to 18.

In the list of prime numbers, the 18th prime number is 61, which when multiplied by the number 18, yields 1,098. The sum of the digits of 1,098 is $1 + 0 + 9 + 8 = 18$.

We can continue this pattern by inserting a 9 between the two digits of 18 and get the following: $198 = 99 + 99$, and again reversing the digits: $891 = 9 \times 99$. While on the topic, we can see that $18 + 81 = 99$, and $9 + 9 = 18$. However, we can extend this strange property as shown here:

| | | | | | |
|---|---|---|---|---|---|
| 18 | = | 9+9 | 81 | = | 9×9 |
| 198 | = | 99+99 | 891 | = | 9×99 |
| 1998 | = | 999+999 | 8991 | = | 9×999 |
| 19998 | = | 9999+9999 | 89991 | = | 9×9999 |

**Equation 4**

...and so it continues!

Another curious property about the number 18 is seen when we take 18 to the third and fourth powers and inspect the results. We will find that each of the digits from 0 to 9 was used exactly once when considering the two results: $18^3 = 5,832$, and $18^4 = 104,976$.

While we are on the 18th prime number, 61, this time we will add it to 18! to get 6,402,373,705,728,061, which just happens to be a prime number.

Similarly, if we take the 18th prime number 61 again, and subtract it from 18!, we get 6,402,373,705,727,939, which is also prime number.

We can also look at the 18th Fibonacci number,[26] and show that it is equal to the sum of the cubes of four consecutive numbers as shown here: $F_{18} = 2,584 = 7^3 + 8^3 + 9^3 + 10^3$.

## THE NUMBER 19

Most women will recall that the Nineteenth Amendment to the Constitution of the United States gave American women the right to vote and was passed by the Congress on June 4, 1919. Also, the 19th state to join the United States was Indiana on December 11, 1816. There are times when the number 19 comes up in curious fashion such as the music composer Bela Bartok finished his Opus 19 in the year 1919 when his age was twice 19 or 38. The number 19 times its reversal 91 yields 1,729, which is the famous Hardy–Ramanujan number (see 1,729 on page 185).

Following are ways the prime number 19 relates to other primes:

- If one takes the sum of the digits of the number 19 and places this sum between the two digits to get the number 1109, a prime number results.
- The number 19 is the smallest prime number that is equal to the sum of the product of its digits and the sum of its digits, as we see: $19 = (1 \times 9) + (1 + 9)$.
- There are a number of ways that the number 19 will generate prime numbers using itself as a base:
  $19^{19} + 19 - 1 = 1,978,419,655,660,313,589,123,997$ is a prime number.
  $19^{19} - 2 = 1,978,419,655,660,313,589,123,977$ is a prime number.
  $2^{19} - 19 = 524,269$ is a prime number.
- The number 19 lends itself to some cute arrangements, such as that $19 - 1$ is a factor of $19^{19} - 1$. Therefore, $\dfrac{19^{19} - 1}{19 - 1} = \dfrac{1978419655660313589123979 - 1}{19 - 1} = \dfrac{1978419655660313589123978}{18} = 109,912,203,092,239,643,840,221$, which is a prime number.
- Analogously, we can also get a prime number from the following calculation: $\dfrac{19^{19} - 2^{19}}{19 - 2} = \dfrac{1978419655660313588599691}{17} = 116377626803547858152923$, which is a prime number.
- The number 19 is the smallest prime number that contains digits that are not prime numbers.

- The number 19 is the smallest prime number whose reversal, 91, is not a prime number.
- The number 19 is the only prime number that is equal to the difference of two cubed prime, namely, $3^3 - 2^3 = 19$.
- The number 19 is the smallest prime, where the sum of the digits of 19 and $19^2 = 361$ are each equal to 10.
- Here we show how we can use 19 ones to create a prime number:
  $$1,111 \times 111^{11} + 111,111,111 + 1 = 35,016,023,542,301,708,082,592,033,$$
  which is a prime number.
- There are times when a nifty pattern up to the number 19 can unexpectedly generate a prime number as we see with the following, where the bases are the initial prime number up to 19 and the exponents are the consecutive natural numbers: $2^1 + 3^2 + 5^3 + 7^4 + 11^5 + 13^6 + 17^7 + 19^8 = 17,398,892,111$, which is a prime number.
- The factorial operation also produces a nice relationship for the number 19, in that $4! - 3! + 2! - 1! = 19$. This process of alternating consecutive factorials when beginning with certain numbers, such as the numbers 3, 4, 5, 6, 7, 8, 10, 15, 41, 59, 61, 105, and 160, as well as the number 19 lead to a prime number. When we begin with the number 19, we get $19! - 18! + 17! - 16! + 15! - 14! + \ldots + 3! - 2! + 1! = 15,578,717,622,022,981$, which is a prime number.
- There exist only five prime numbers less than 180, namely, 2, 3, 7, 19, and 31, where the relationship $\dfrac{n^n - 1}{n - 1}$ for $n$ leads to another prime number. As we have shown earlier, for the number 19, we have $\dfrac{19^{19} - 1}{19 - 1} = \dfrac{1,978,419,655,660,313,589,123,978}{18} = 109,912,203,092,239,643,840,221$, which is a prime number. The reader may wish to apply this relationship to the remaining four numbers for which this also holds true.

As the number 19 is raised to various powers, compare the list of exponents to the resulting sum:

$19^5 + 19^2 + 19^1 + 19^3 + 19^5 + 19^6 + 19^4 + 19^0 = 52,135,640$. Notice the relationship of the digits!

One often neglected curiosity is that the digit 8 appears 19 times in the numbers from 1 to 100.

Another curiosity involving the number 19, where the sum of the first 19 numbers is 10 times the number 19, as shown here: $19 + 18 + 17 + 16 + 15 + 14 + 13 + 12 + 11 + 10 + 9 + 8 + 7 + 6 + 5 + 4 + 3 + 2 + 1 + 0 = 190$.

Using the first 19 triangular numbers as denominators for unit fractions and multiplying that sum by 10 will result, surprisingly, in the number 19, as we show here:

$$10\left(\frac{1}{1}+\frac{1}{3}+\frac{1}{6}+\frac{1}{10}+\frac{1}{15}+\frac{1}{21}+\frac{1}{28}+\frac{1}{36}+\frac{1}{45}+\frac{1}{55}+\frac{1}{66}+\frac{1}{78}+\frac{1}{91}+\right.$$

$$\left.\frac{1}{105}+\frac{1}{120}+\frac{1}{136}+\frac{1}{153}+\frac{1}{171}+\frac{1}{190}\right)=19$$

While on triangular numbers the number 19 squared can be written as a sum of three consecutive triangular numbers squared: $6^2 + 10^2 + 15^2 = 36 + 100 + 225 = 361 = 19^2$.

There is an interesting test for divisibility by 19. Take the number that you are checking for divisibility by 19, delete the last digit, and then add double of this digit to the remaining number. Continue this process until a number is reached that you can visually determine if it is divisible by 19. If the end result is divisible by 19, then the original number was divisible by 19. For example, suppose we want to determine whether the number 1,387 is divisible by 19. We first delete the last digit, 7, and add twice that digit from the remaining number: $138 + 14 = 152$. Repeating this process $15 + 4 = 19$, therefore, the original number 1,387 is divisible by 19.

Another example where we can use the number 67,982. Follow along the various steps:

$6,798 + 4 = 6,802; 680 + 4 = 684; 68 + 8 = 76; 7 + 12 = 19,$     which verifies that the original number 67,982 is divisible by 19.

There is another interesting test for divisibility by 19, where we would have to represent the original number in the form of $100p + q$, and if, and only if, $p + 4q$ is divisible by 19 then the original numbers are divisible by 19. Checking once again the number 1,387 for divisibility by 19, we would represent that number as $(13 \times 100) + 87$. We now notice that $13 + (4 \times 87) = 361$, which happens to be $19^2$, and so it is divisible by 19; therefore, the original number 1,387 is divisible by 19.

Although these divisibility tests are not particularly efficient, they do give an insight into numbers, in this case the number 19. Another procedure for determining whether a given number is divisible by 19 is to delete the units digit and multiply it by 17 and then subtracting that number from the remainder of the original number. Continue this process until a number results that can be visually determined if it is divisible by 19 or not. For example, consider the number 34,542. To test divisibility by 19 using this latest procedure, we delete the units digit, 2, and subtract $17 \times 2 = 34$ from the remaining number 3,454, to get 3,420. Following this procedure, the resulting number is 342. Once again, deleting the units digit and multiplying it by 17 to get 34 and subtracting that from the remaining number, 34, we get 0, which is a multiple of 19, and therefore, the original number 34,542 is divisible by 19.

Yet another test for divisibility by 19, which is a bit more cumbersome but can be effective in demonstrating the power of numbers (not necessarily more effective than a calculator). The process goes as follows: delete the units digit from the number you are testing and multiply the remaining number by 9, after which you subtract the units digit. Continue this process until you reach a number that you can visually recognize as either zero or a multiple of 19. For example, supposing we want to determine if the number 133 is divisible by 19. We follow the process:

$(13 \times 9) - 3 = 114$, then $(11 \times 9) - 4 = 95$, then $(9 \times 9) - 5 = 76$, then $(7 \times 9) - 6 = 57$, then $(5 \times 9) - 7 = 38$, then $(3 \times 9) - 8 = 19$. Therefore, the number 133 is divisible by 19. Naturally, one could have stopped much sooner, once the number arrived at was determined to be divisible by 19.

## THE NUMBER 20

Perhaps because the number 20 is twice the base 10, we refer to the number 20 as a "score," most likely taken from the old English word "scoru" meaning 20. One of the most famous speeches in American history was that given by President Abraham Lincoln in Gettysburg, Pennsylvania on November 19, 1863. The first line of this address reads "Four score and seven years ago our fathers brought forth on this continent a new nation, " It should also be noted that the state of Mississippi was the 20th state to join the United States on December 10, 1817. When one has vision checked and the result comes back as normal vision, it is said to be a 20/20 vision. There is a curious connection in temperature gauges, where there is a direct integer connection between Celsius and Fahrenheit. When the temperature reads 20° Celsius is equal to 68° Fahrenheit, which is the recommended indoor temperature during the winter.

In geometry the number 20 presents itself among the Platonic solids, where an icosahedron has 20 faces and a dodecahedron as 20 vertices.

In previous number presentations we discussed the nature of triangular numbers with the number 20. We have reached a number, which is the sum of the first 4 triangular numbers, namely, $20 = 1 + 3 + 6 + 10$.

The 20th prime number 71, and it is a factor of the sum of the first 20 primes: $2 + 3 + 5 + 7 + 11 + 13 + 17 + 19 + 23 + 29 + 31 + 37 + 41 + 43 + 47 + 53 + 59 + 61 + 67 + 71 = 639$ because $9 \times 71 = 639$.

The number 20 is the smallest number that cannot be made into a prime number by placing a digit before it or after it. Certainly, by placing any digit before it the number would be divisible by 10 when you place any of the nine digits after the number 20, such as: 201, 202, 203, 204, 205, 206, 207, 208, and 209, which are all composite numbers.

## THE NUMBER 21

To many Americans what may come to mind with the number 21 is the card game that carries that name and also sometimes carries the name "blackjack." Some may know that the Twenty-First Amendment to the Constitution repealed the Eighteenth Amendment and allowed Americans to once again drink alcoholic beverages. The state of Illinois was the 21st state to join the United States on December 3, 1818. In many countries at age 21 a person becomes an adult and is able to carry the privileges of adulthood such as voting, purchasing adult products, gambling, and so forth. Yet the number 21 has other appearances in culture such as perhaps the name of some films, books, and art. Of course, in the United States, the president and the national flag are honored with a "21-gun salute."

Arithmetically, the number 21 does provide us with a most unusual method of multiplication. To multiply a number by 21 you merely can double the number being multiplied by 21, and then multiply the resulting number by 10, and finally add the original number. Perhaps this can be best seen through an example. Supposing we would like to multiply 37 by 21. We begin by doubling the number 37, which yields 74, and then multiplying the result by 10 to get 740, and finally we add the original number 37 to get 777, which is equal to $37 \times 21$.

In mathematics we introduce the number 21 by highlighting the fact that it is the sixth triangular number, as we can see in Figure 43. This also means that the number can be represented as $1 + 2 + 3 + 4 + 5 + 6 = 21$.

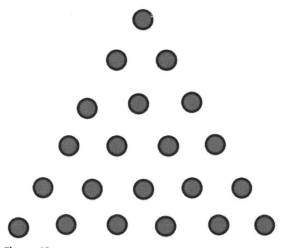

**Figure 43**

Among the famous Fibonacci numbers, the number 21 is the only member of the sequence that is the sum of three distinct Fibonacci primes, namely: $21 = 3 + 5 + 13$.

The number 21 also represents the sum of the numbers (dots) on a single die, shown in Figure 44.

**Figure 44**

## THE NUMBER 22

In literature we are familiar with Joseph Heller's (1923–1999) novel *Catch–22*, which has also been popularized in 1970 through the film of the same name, as it gave rise to the conundrum that one encounters because of contradictory conditions faced. The 22nd state to join the United State was Alabama on December 14, 1819.

Unlike its predecessor, the number 22 has quite a few curiosities in mathematics such as:

- We notice that 22 is clearly a palindromic number because it reads in both directions the same. Yet, the number $22^2 = 484$ is also a palindromic number.
- The number 22 also has another distinction in that it is considered a semiprime, which is a number that is the product of two primes, namely, in this case 2 and 11.
- Another curiosity about the number 22 is that the sum of its digits, 4, is equal to the sum of the digits of its prime factors, namely 2 and 11.
- The number 22 is the smallest number that can be expressed as the sum of two primes in three different ways: $22 = 3 + 19 = 5 + 17 = 11 + 11$.
- The number 22 is also a pentagonal number as we can see in Figure 45.

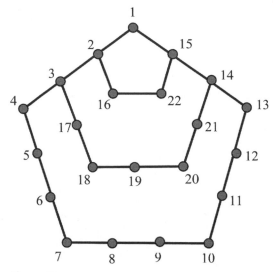

**Figure 45**

The number 22 also shows itself very interestingly in geometry because if you take six lines cutting through a circle with the maximum number of intersections, that is, no three lines concurrent, the circle will be partitioned into 22 parts as you can see in Figure 46.

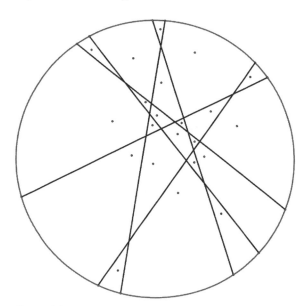

**Figure 46**

## THE NUMBER 23

To many people number 23 reminds them of the 23rd Psalm from the Book of Psalms. In the King James version of the New Testament the word "hell" is found 23 times. The number 23 also represents a number of chromosome pairs in the human body, which, by the way, also has 23 discs in the spine. We should not forget that the 23rd state to join the United States was the state of Maine on March 15, 1820.

The number 23 probably has more significance to the broader audience in mathematics. It is clear that the number 23 is a prime number; however, it also has the distinction as being the smallest odd-prime number that is *not* a member of a twin-prime pair (two prime numbers whose difference is 2). In other words, the previous prime, 19, and the succeeding prime, 29, are each more than two numbers apart from 23.

The sum of the first three primes, $2 + 3 + 5 = 10$, the next power of 10, which is 100, can be achieved by taking the consecutive primes up to the number 23, such as: $2 + 3 + 5 + 7 + 11 + 13 + 17 + 19 + 23 = 100$. It is not known whether any other consecutive listing of primes will lead to a higher power of 10.

In mathematics peculiar things can happen unexpectedly. It is simple to experience that if $n$ takes on most values in $\dfrac{10^n - 1}{9}$ the result will be a repunit number, namely a number with only 1s, such as when $n = 3$, $\dfrac{10^3 - 1}{9} = \dfrac{1000 - 1}{9} = 111$, which is a composite number as $111 = 37 \times 3$. However, to date it has been found that only five values for $n$ yield a prime number. That is where we find the number 23 in a special place because, for $n = 23$, we find that $\dfrac{10^{23} - 1}{9} = 11{,}111{,}111{,}111{,}111{,}111{,}111{,}111{,}111$, which is a prime number! This is a rather rare application, where the number 23 shares this unusual feature with the numbers 2, 19, 317, and 1,031.

Furthermore, using 23 ones, that is, a 23 repunit, we can generate a prime number as follows:

$11{,}111 \times 1{,}111^{11} + 11{,}111{,}111{,}111 + 1 = 35{,}367{,}779{,}633{,}655{,}045{,}365{,}782{,}$
$864{,}884{,}274{,}523{,}033$, which is a prime number.

In the history of mathematics, female mathematicians were a rarity centuries ago. However, the famous French mathematician Sophie Germain (1776–1831) is honored by having a prime number named after her. Prime numbers, $p$, are those primes where $2p + 1$ is also a prime number. The first *Sophie Germain prime number* is the number 3 because $(2 \times 3) + 1 = 7$ is a prime number. The number 23 is the fifth Sophie Germain prime number because $(2 \times 23) + 1 = 47$, which is a prime number.

The number 23 is the smallest prime number where the sum of the squares of its digits, $2^2 + 3^2$, is also a prime number, namely, 13.

Here are a few cute arrangements with the digits of 23:

- An unexpected symmetry for the number 23 is:
  $$0^5 + 1^4 + 2^3 + 3^2 + 4^1 + 5^0 = 0 + 1 + 8 + 9 + 4 + 1 = 23.$$
- $23 = (2^2 + 3^3) - (2! + 3!)$, that is: $(4 + 27) - (2 + 6) = 23$.
- $23 = -(2^2 - 3^3)$, that means $-(4 - 27) = 23$.
- Consider the first three prime numbers, 2, 3, and 5, and then multiply them consecutively:
  $$(1 \times 2) + (2 \times 3) + (3 \times 5) = 2 + 6 + 15 = 23$$
- Another interesting arrangement that results in 23 is: $1! + (2! + 2!) + (3! + 3! + 3!) = 1 + 4 + 18 = 23$.

- The sum of the first 23 primes is $874 = 23 \times 38$, and the 23rd prime happens to be 83. How is that for a coincidence?
- Suppose we find the sum of the digits of the number $23^{2+3}$, which is 6,436,343. Now take the digit sum to get $6 + 4 + 3 + 6 + 3 + 4 + 3 = 23$. What a surprise!
- The number 23 the smallest prime that is equal to the sum of the sum and product of a pair of twin primes, as we have in the following: $(3 + 5) + (3 \times 5) = 23$.
- The number 23 the only prime number, where its factorial, $23! = 25{,}852{,}016{,}738{,}884{,}976{,}640{,}000$ is a number, which has the number of digits equal to the original number. In other words, in this case 23 digits.
- Here is a curious way to achieve number 23: $\dfrac{3^5 + 5^3}{(3 + 5) + (5 + 3)} = \dfrac{368}{16} = 23.$

A motivated reader may wish to try to take any two-digit prime number and place between the two digits the next prime number in the sequence of prime numbers. The smallest one that yields a four-digit prime number is number 23, with the number 2,**293**.

As a parting curiosity with the number 23, consider the sum of the first 23 prime numbers:

$$2 + 3 + 5 + 7 + 11 + 13 + 17 + 19 + 23 + 29 + 31 + 37 + 41 + 43 + 47$$
$$+ 47 + 53 + 59 + 61 + 67 + 71 + 73 + 79 + 83 = 874, \text{ which is divis-}$$
ible by 23, in that $874 \div 23 = 38$.

We leave the number 23 with a curious coincidence. The 23rd letter of the alphabet is the W and on the QWERTY keyboard it is located exactly beneath the 2 and 3 (see Figure 47).

**Figure 47**

## THE NUMBER 24

Most likely the first thought that comes to mind with the number 24 is the number of hours in the day. To avid Major League Baseball fans, the number 24 has been made famous by one of the most talented baseball players, Willie Mays, whose uniform carried that number. When you have a gold ring or bracelet that is marked with 24 karat gold, that is tantamount to indicating that it is pure gold. The number 24 also indicates the number of letters in the Greek alphabet. By the way, the 24th state to enter the United States was the state of Missouri on August 10, 1821.

In mathematics, the number 24 harbors many peculiarities, such as the simple one, where 24 is divisible by both the sum of its digits (6) and the product of its digits (8). The number 24 is also equal to a factorial, namely 4!

The number 24 is also the smallest number which has exactly eight divisors, namely 1, 2, 3, 4, 6, 8, 12, and 24. If we now subtract 1 from each of these divisors (except the first two) we will end up with six prime numbers.

Furthermore, if we take the product of the proper divisors of 24, a perfect cube results, namely $2 \times 3 \times 4 \times 6 \times 8 \times 12 = 13,824 = 24^3$.

Curiously enough, if we take the sum of the squares of the first 24 numbers, we end up with another square number, as we can see here: $1^2 + 2^2 + 3^2 + 4^2 \ldots 23^2 + 24^2 = 4,900 = 70^2$.

We mentioned earlier that twin primes are two prime numbers that differ by 2. The number 24 can be represented as the sum of two twin primes, namely $11 + 13 = 24$. We can also generate a pair of twin primes with these digits as follows: $24^{2+4} + (2 + 4) + 1 = \mathbf{191,102,983}$, and $24^{2+4} + (2 + 4) - 1 = \mathbf{191,102,981}$. Yet, another way of using 24 to generate a pair of twin primes is: $(24 \times 24) + 2,424 + 1 = 3,001$, and $(24 \times 24) + 2,424 - 1 = 2999$, which are the 431st and 430th prime numbers, respectively.

The number 24 is the smallest number that can be expressed as the sum of two prime number in three ways, as follows: $24 = 5 + 19 = 7 + 17 = 11 + 13$.

As we move along, we see that the number 24 has a unique relationship with the number 26. If we take the sum of the squares of the divisors of 24: $1^2 + 2^2 + 3^2 + 4^2 + 6^2 + 8^2 + 12^2 + 24^2 = 850$, and now, the sum of the squares of the divisors of 26: $1^2 + 2^2 + 13^2 + 26^2 = 850$. This is a good lead-in to the number 26, coming soon!

## THE NUMBER 25

Because the number 25 is one-quarter of 100, the US currency has a coin entitled "quarter" as it is worth \$0.25 or one-quarter of a dollar. The state of Arkansas was the 25th state to join the United States on June 15, 1836.

The famous French mathematician, Pierre de Fermat (1607–1655) claimed that the only square number that was 2 less than a cube is the number 25 as $5^2 = 3^3 - 2$.

The number 25 is also the sum of all the odd single-digit numbers, namely $1 + 3 + 5 + 7 + 9 = 25$. Furthermore, the number 25 is the only prime square number whose digits are all prime numbers.

Another curiosity about the number 25 is that every power of 25 has the tens and units digit as 25; for example, $25^2 = 625$; $25^3 = 15,625$; $25^4 = 390,625$; $25^5 = 9,765,625$; and $25^6 = 244,140,625$. This leads us to even more surprising result that all the powers of the number 25 not only ending 25, but also in 625.

As an odd perfect square, the number 25 can be represented as the sum of two consecutive integers, as $25 = 12 + 13$. A symmetric pattern that leads to the number 25 is as follows:

$1 + 2 + 3 + 4 + 5 + 4 + 3 + 2 + 1 = 25$.

Along with the numbers 121 and 5,041, the number 25 can be expressed as a factorial +1, as $4! + 1 = 25$.

## THE NUMBER 26

When thinking of the English alphabet, the number 26 comes up readily as it represents the number of letters in the alphabet. On January 26, 1837, the state of Michigan was the 26th to join the United States.

The number 26 does not enjoy many curiosities, especially because it is placed between a perfect square $5^2 = 25$, and a perfect cube $9^3 = 27$. We would have to go to $26^2 = 676$ to get something special, in this case, a palindromic number. Yet, if we look at $26^3 = 17,576$, and then take the sum of these digits: $1 + 7 + 5 + 7 + 6$, we get 26.

Among the first 100 prime numbers there are 26 numbers with a units-digit of 3, which is more than the number of prime numbers ending in 1, 7, or 9. While considering prime numbers, it also should be noted that there are no primes between $26^2$ and $28^2$.

Another curious aspect occurs when we add or subtract the number 123,456,789 with $10^{26}$.

In the first case, $123,456,789 + 10^{26}$, we get the number 100,000,000, 000,000,000,123,456,789, and in the second case, $10^{26} - 123,456,789$, we get 99,999,999,999,999,999,876,543,211, the amazing thing is that both of these numbers are prime numbers. Notice the patterns of these two large numbers and then appreciate their sum: 200,000,000,000,000,000,000,000,000.

## THE NUMBER 27

Unlike the English alphabet, the Spanish alphabet has 27 letters because it has the letter Ñ in addition to the 26 letters of the English alphabet. Florida was the 27th state to join the United States on March 3, 1845.

The number 27 is the smallest composite number that cannot be represented as the sum of two prime numbers. We can see that 27 is the only positive integer that is three times the sum of its digits, 9, or $3 \times 9 = 27$.

Another cute pattern for the number 27 is if you add the consecutive numbers beginning and ending with its digits 2 and 7, you will get $2 + 3 + 4 + 5 + 6 + 7 = 27$.

Another curiosity to consider is that if we take the sum of the digits of $27^3 = 19,683$ we get $1 + 9 + 6 + 8 + 3 = 27$. Furthermore, the number 27 is the smallest number which can be expressed as a sum of 3 squares in two different ways: $3^2 + 3^2 + 3^2 = 27 = 5^2 + 1^2 + 1^2$.

Number 27 does have a rather unique place amongst two-digit numbers (which it shares with the number 37). This is that every three-digit multiple of 27, such as $27 \times 13 = 351$, when cyclically altered as perhaps the various cyclical alternatives, 513 and 135, are each a multiple of 27, and $19 \times 27 = 351$, and $5 \times 27 = 135$.

The number 27 is the smallest composite number $n$ greater than 1, where $n! + 1$ is a prime number, namely, $27! + 1 = 10,888,869,450,418,352,160,768,$ $000,001$, which is, in fact, a prime number.

We can also generate a prime number using the number 27 as follows: $27^{27} + (2 \times 7)^{(2 \times 7)} = 443,426,488,243,037,769,948,249,630,619,149,892,803$ $+ 11,112,006,825,558,016 = 443,426,488,243,037,769,948,260,742,625,975,$ $450,819$, which is a prime number.

We find that the decimal equivalent to $\frac{1}{7}$ = 0.142857 142857 142857 142857 142857..., is analogously similar to the other fractions with denominator 7. If we consider the repeating parts of each of the first six fractions with denominator 7 as shown in Figure 48, we have a "somewhat magic square"—rows and columns, but not diagonals, have the same sum. All the rows and all the columns have a sum of 27, which gives this number a special place amongst this curious form of a magic square.

$$\frac{1}{7} = 1\ 4\ 2\ 8\ 5\ 7$$

$$\frac{2}{7} = 2\ 8\ 5\ 7\ 1\ 4$$

$$\frac{3}{7} = 4\ 2\ 8\ 5\ 7\ 1$$

$$\frac{4}{7} = 5\ 7\ 1\ 4\ 2\ 8$$

$$\frac{5}{7} = 7\ 1\ 4\ 2\ 8\ 5$$

$$\frac{6}{7} = 8\ 5\ 7\ 1\ 4\ 2$$

**Figure 48**

## THE NUMBER 28

First reckoning of the number 28 is probably that there are 28 days in a lunar month. Game enthusiasts will surely know that there are 28 pieces in a traditional set in the game of Dominoes. On December 29, 1845, the state of Texas became the 28th state to join the United States.

With the number 28 there are two very distinct characteristics. First, it is the second *perfect number* (after the perfect number 6) because the sum of its factors: 1 + 2 + 4 + 7 + 14 is equal to 28. The next few perfect numbers after the number 28 are:

496
8,128
33,550,336
8,589,869,056
137,438,691,328

Notice how the perfect numbers all end in either the digits 6 or 8. As with other perfect numbers, if we add the number 28 to the sum of its factors to then get the sum $1 + 2 + 4 + 7 + 14 + 28$ and then take the reciprocals of these numbers, we get $\dfrac{1}{1} + \dfrac{1}{2} + \dfrac{1}{4} + \dfrac{1}{7} + \dfrac{1}{14} + \dfrac{1}{28} = 2$, which is true for all perfect numbers. Another rather interesting feature of the number 28 is that it is considered a *harmonic divisor number*, which is a positive integer whose divisors have an integer harmonic mean. A harmonic mean is defined as the reciprocal of the average (arithmetic mean) of the reciprocals of a set of numbers and can be found for the five divisors of 28 to be equal to

$$\cfrac{1}{\cfrac{\dfrac{1}{1} + \dfrac{1}{2} + \dfrac{1}{4} + \dfrac{1}{7} + \dfrac{1}{14}}{5}} = \frac{5(1 \times 2 \times 4 \times 7 \times 14)}{1 + 2 + 4 + 7 + 14} = \frac{3920}{28} = 140,$$ which is a

positive integer and, therefore, number 28 is considered a harmonic divisor number.

From an arithmetical perspective the number 28 has a very unusual characteristic, that is,

- It can be expressed as a sum of the first nonnegative integers: $1 + 2 + 3 + 4 + 5 + 6 + 7 = 28$, and
- It can be expressed as a sum of the first primes: $2 + 3 + 5 + 7 + 11 = 28$, and
- It can be expressed as the first nonprime numbers: $1 + 4 + 6 + 8 + 9 = 28$.
- Also, $28 = 1^3 + 3^3$.

There is a nice symmetry with the number 28:

$$\frac{28! + 1}{28 + 1} = \frac{304,888,344,611,713,860,501,504,000,001}{29}$$

$$= 10,513,391,193,507,374,500,051,862,069,$$ which is a prime number with $28 + 1$ digits.

Another interesting generation of a prime number using the number 28 is as follows:

$$28! + 28^{28} + 1 = 304,888,344,611,713,860,501,504,000,000 +$$
$$33,145,523,113,253,374,862,572,728,253,364,605,812,736 + 1 =$$
$$33,145,523,113,558,263,207,184,442,113,866,109,812,737,$$ which is a prime number.

Additionally, the number 28 is also a triangular number because $1 + 2 + 3 + 4 + 5 + 6 + 7 = 28$. We can also exhibit the triangular number geometrically with the number of dots shown in Figure 49.

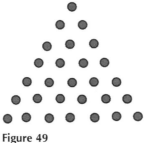

**Figure 49**

The number 28 is also a hexagonal number as shown geometrically in Figure 50.

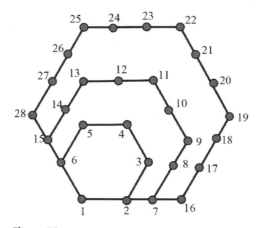

**Figure 50**

It is also a *Sophie Germain prime*, which was mentioned earlier as a prime number *n* where a number of the form $2n + 1$ is also a prime number. In this case, $(2 \times 28) + 1 = 57$, which is a prime, thus making 28 a Sophie Germain prime.

## THE NUMBER 29

On the calendar, the number 29 has a special place in the month of February, as it appears every four years as a leap year date. On a larger scope, the planet Saturn orbits the sun every 29 years. Another fact about the number 29 is that the human skull consists of 29 bones. On December 28, 1846, Iowa was the 29th state to join the United States. While considering states, it should be noted that all ZIP codes in the state of South Carolina begin with the number 29.

The number 29 is the only two-digit prime number, where exactly one digit is even and one digit is a composite odd. The number 29 is also the smallest prime number when placed between the sum of its digits, 11, the number 1,291 is created, which is the 210th prime number.

The number 29 is a prime number, and a member of a twin pair of primes, namely, 29 and 31.

In 1798, the French mathematician Adrien-Marie Legendre (1752–1833) showed that for values of $n$ from 0 to 28 the relationship $2n^2 + 29$ will yield 29 distinct prime numbers, such as when $n = 1$, prime number 31 is obtained, and when $n = 5$, the prime number 79 is reached.

At a quick glance, we notice that 29 is the smallest prime number where the sum of the digits $(2 + 9 = 11)$ and the difference of the digits $(9 - 2 = 7)$ are both prime numbers. Furthermore, the prime number following the 29th prime is the $(2 + 9)^{th}$ prime, namely 31.

As we search further for curiosities involving the number 29, we find that $2^2 + 3^2 + 4^2 = 29$, making 29 the smallest prime number, which is equal to the sum of three consecutive squares. And further, the number 29 is a smallest prime that can be expressed in terms of four consecutive prime numbers as follows: $(2 \times 7) + (3 \times 5)$.

A motivated reader might have noticed that when you add the number 29 to its reverse 92, you get 121 which is a perfect square. While on reversals, we note that 29 is a smallest prime number whose reversal, 92, is a sum of the reversals of two double-digit prime numbers: $31 + 61 = 92$. While dealing with the number 29 in its reverse 92, we can also generate a prime number as follows:

$2^9 + 9^2 = 512 + 81 = 593$, which is the 108th prime number. Furthermore, just imagine that the reversal of 29 is divisible by the previous prime number, 23. In other words, $92 \div 23 = 4$.

The number 29 is the last number in a series of consecutive squares, whose sum is a square number and whose square root happens to be the reverse of 29, as shown here:

$$7^2 + 8^2 + 9^2 + \ldots + 27^2 + 28^2 + 29^2 = 8,464 = 92^2.$$

Another curiosity is that $2^{29} = 536,870,912$, which is the largest power of 2, where all the digits are distinct. We can see this because the next power of 2, which is $2^{30} = 1,073,741,824$, already has some repeating digits; clearly any higher power of 2 will generate numbers of more than 10 digits, where a repeating digit is unavoidable. Curiously, the number 29! is a smallest factorial in which each digit appears at least twice, as we can see $29! = 8,841,761,993,739,701,954,543,616,000,000$. The number 29 is the largest power of 5 that contains no to consecutive equal digits. That is, $5^{29} = 18,626,451,492,309,570,3125$. A motivated reader may want to check higher powers of 5 to notice that there are at least two like-digits together, such as with $5^{30} = 931,322,574,615,478,515,625$, where a pair of 2s are together.

For a nice little symmetry, we find that $2^{29} + 1$ is divisible by $2(29) + 1$. That is, $\dfrac{2^{29} + 1}{2(29) + 1} = 9{,}099{,}507$.

The number 29 is the smallest multidigit prime number with a product of the digits of $29^3 = 24{,}389$ is also a perfect cube, as we can see with $2 \times 4 \times 3 \times 8 \times 9 = 1{,}728 = 12^3$.

The number 29 is the smallest two-digit prime number, which is equal to the sum of the digits of all the previous two-digit primes. In other words, the sum of the digits of 11, 13, 17, 19, and 23, whose digit-sum is $1 + 1 + 1 + 3 + 1 + 7 + 1 + 9 + 2 + 3 = 29$.

The number 29 squared is the only two-digit prime number that is the sum of the squares of two consecutive two-digit numbers. That is, $20^2 + 21^2 = 400 + 441 = 841 = 29^2$.

We can set up an arithmetic progression of prime numbers ending with 29, where the common difference is 6, as we can see with the following: 5, 11, 17, 23, 29. This is the smallest listing of five prime numbers in an arithmetic progression.

In some strange ways, the number 29 helps to generate a prime number as follows: $3^{29} - 2^{29} = 68{,}630{,}377{,}364{,}883 - 536{,}870{,}912 = 68{,}629{,}840{,}493{,}971$, which is a prime number.

Another oddity is that after the numbers 1 and 5, the number 29 has an unusual characteristic that
$$2 \times 29^2 - 1 = 1682 - 1 = 41^2.$$
We leave the number 29 as there is a unique set of 29 prime numbers, each of which consists of the five odd digits. They are 13,597, 13,759, 15,739, 15,937, 15,973, 17,359, 17,539, 19,753, 39,157, 51,973, 53,197, 53,719, 53,917, 57,139, 57,193, 71,359, 71,593, 73,951, 75,193, 75,391, 75,913, 75,931, 79,153, 79,531, 91,573, 91,753, 95,317, 95,713, 95,731. This is quite astonishing!

## THE NUMBER 30

In Europe, the 30-Years War which lasted from 1618 to 1648 took many lives in central Europe and Germany. The number 30 represents the number of letters in the Bulgarian and Serbian alphabets. There are currently 30 teams in the National Basketball Association and there are 30 teams in Major League Baseball. The state of Wisconsin was the 30th state to join the United States on May 29, 1848. According to the Bible (2 Samuel 5:4) King David was 30 years old when he began his reign.

The number 30 is the smallest number that has three distinct prime factors, namely, 2, 3, and 5. If we take the reciprocals of these factors and subtract the reciprocal of 30, we, surprisingly, get the following: $\dfrac{1}{2} + \dfrac{1}{3} + \dfrac{1}{5} - \dfrac{1}{30} = 1$. Furthermore, the number 30 is also the smallest number that cannot be expressed as a sum of three cubes.

The number 30 can be shown to be the sum of the first four square numbers as we see here:

$$1^2 + 2^2 + 3^2 + 4^2 = 1 + 4 + 9 + 16 = 30.$$

The number 30 leads an arithmetic progression of numbers, each of which contains exactly three distinct prime factors. This arithmetic progression is 30, 66, 102, 138. Specifically, the number 66 has as its prime factors 2, 3, and 11, and the number 102 has as its prime factors 2, 3, and 17, and the number 138 has as its prime factors the numbers 2, 3, and 23.

There are number of ways that we can generate prime numbers using the number 30:

$30 \times 2^{30} - 1 = 32,212,254,719$, which is a prime number.

$30^{30} + 30 - 1 = 205891,132,094,649,000,000,000,000,000,$
$\qquad\qquad 000,000,000,029,$ which is a prime number.

The number 30 is the smallest number that fits the following scheme: $30^{30+2} + 1$ is a prime number, namely,

185,302,018,885,184,100,000,000,000,000,000,000,000,000,001.

Geometrically, the number 30 manifests itself in the number of edges, which both a dodecahedron and icosahedron have, as can be seen in Figure 51.

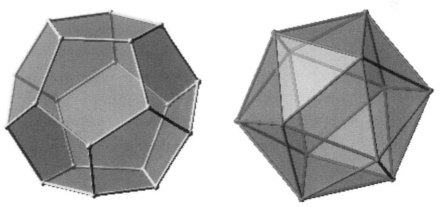

**Figure 51**

## THE NUMBER 31

For Californians, the number 31 has a specific significance since on September 9, 1850, California became the 31st state of the United States. And by the way, the word *California* begins and ends with the 3rd and 1st letter of the alphabet, respectively, just to further demonstrate its allegiance to the number 31. In 1953, the Baskin-Robbins chain of stores highlighted the number 31 on its logo to indicate its variety of ice creams. However, it is also notable that 31 is the largest prime factor of 1,953, which equals $31 \times 7 \times 3 \times 3$.

A "fun" aspect of the number 31 is found by squaring the number 31 and flipping it over to get another perfect square, as $31^2 = 961$. Flipping 961 over to get $196 = 14^2$. Here is another cute relationship with the number 31: see how the number 31 is placed in parts here:

$$3! + 1! + (3 + 1)! = (2 \times 3) + 1 + (2 \times 3 \times 4) = 31.$$

A quick way to multiply by 31 could be to triple the number being multiplied by 31, and then multiplying the result by 10 and finally adding the original number. Perhaps we can best see this with the following example. Suppose we wish to multiply 43 by 31. The first step is to triple 43, which yields 129, and then multiply this result by 10 to get 1,290, and finally we add the original number 43 to get 1,333, which turns out to be $43 \times 31$.

The prime numbers leading up to number 31 are: 2, 3, 5, 7, 11, 13, 17, 19, 23, 29, 31. When we take the sum of the digits of these prime numbers, we get:
$2 + 3 + 5 + 7 + 1 + 1 + 1 + 3 + 1 + 7 + 1 + 9 + 2 + 3 + 2 + 9 + 3 + 1 = 61,$
which is a prime number.

For the most part, odd numbers seem to give us some truly interesting curiosities. The number 31 is clearly a prime number, which by now should be easily recognizable. As was mentioned earlier, the number 31 is the second member of a twin pair of prime numbers coupled with the number 29. We also should note that $31^7 = 27,512,614,111$, where the sum of the digits is:
$2 + 7 + 5 + 1 + 2 + 6 + 1 + 4 + 1 + 1 + 1 = 31.$

The number 31 can also be expressed in some interesting ways, for example: $31 = 2^0 + 2^1 + 2^2 + 2^3 + 2^4$, and $31 = 5^0 + 5^1 + 5^2$. The only number other than 31 and with 1 as the units digit that can be expressed as a sum of successive powers is the number 8,191 (see page 190).

Powers of 31 can also lead to prime numbers as shown in the following cases:

$$\frac{31^{31} - 1}{31 - 1} = \frac{17,069,174,130,723,235,958,610,643,029,059,314,756,044,734,430}{30}$$

$$= 568,972,471,024,107,865,287,021,434,301,977,158,534,824,481,$$

which is a prime number.

Also, $3^{31} - 2^{31} = 617,673,396,283,947 - 2,147,483,648 = 617,671,248,$ $800,299$ is a prime number.

Even the reciprocal of the number 31 provides us with a curious result. Consider the following expansion: $\frac{1}{31} = 0.\overline{03225\ 80645\ 16129}$. These 15 digits repeat indefinitely. The number 31 is the first number whose reciprocal yields an odd number of digits, 15, in its decimal expansion before it repeats. Even this decimal expansion provides us with another unexpected curiosity, namely that the sum of the numbers in the repeated decimal expansion when seen as three five-digit numbers: $03,225 + 80,645 + 16,129 = 99,999$. If we break this decimal expansion into other small parts such as the following: $032 + 258 + 065 + 416 + 129 = 900$.

There is still another rather unusual occurrence here with the first five digits of the expansion multiplied as follows:

| | |
|---|---|
| $032258 \times 2 = 64516$ | $032258 \times 9 = 290322$ |
| $032258 \times 4 = 129032$ | $032258 \times 14 = 451612$ |
| $032258 \times 5 = 161290$ | $032258 \times 16 = 516128$ |
| $032258 \times 7 = 225806$ | $032258 \times 18 = 580644$ |
| $032258 \times 8 = 258064$ | $032258 \times 19 = 612902$ |

**Equation 5**

When you search for these products of six digits each in the decimal expansion of the reciprocal of 31, you should be able to locate them somewhere in the stream of the decimal expansion: 0322580645161290322580645116129.... For example, $032,258 \times 8 = 258,064$ can be found in the sequence of numbers is highlighted here: 03**225806451**61290322580645116129....

Another curiosity involving the number 31 can be seen with two partner fractions that both yield prime numbers: $\frac{2^{31} - 1}{2 - 1} = 2,147,483,647$, which is a prime number; and $\frac{2^{31} + 1}{2 + 1} = 715,827,883$, which is also a prime number.

The number 31 is the lead number for a sequence of curious looking numbers that are all primes: 31, 331, 3,331, 33331, 333,331, 3,333,331, 33,333,331, while the next ones are no longer prime numbers since $333,333,331 = 17 \times 19,607,843$, and $3,333,333,331 = 673 \times 4,952,947$, and so on.

Here is one that is truly amazing. Consider the number 31,111,111,111, 111,111,111,111,111,111,111,111, which has 31 units digits. If we divide that number by 31, we get 1,003,584,229,390,681,003,584,229,390,681, which is a prime number.

Recall the famous Fibonacci numbers: $1, 1, 2, 3, 5, 8, 13, 21, 34, \ldots$, where the 31st Fibonacci number is 1,346,269 and the sum of its digits is 31.

We even have happy numbers in mathematics, which are numbers, where summing the squares of the digits and repeating this process will lead to the number 1. The number 31 is a happy number and is the first of a pair of happy numbers, namely 31 and 32. That is:

$3^2 + 1^2 = 10$ and $1^2 + 0^2 = 1$. Also, the next number 32 is also a happy number because $3^2 + 2^2 = 13$, and $1^2 + 3^2 = 10$ and $1^2 + 0^2 = 1$.

The number 31 is the smallest prime number that can be expressed as a sum of two triangular numbers in two different ways: $31 = 3 + 28$ and $31 = 10 + 21$. The number 31 can also be expressed as a sum of three consecutive prime numbers as: $31 = 7 + 11 + 13$.

Consider the following sequence and find the next number: **1, 2, 4, 8, 16**. Clearly, most people would guess that 32 is the next number. Yes, that would be fine. However, when the next number, surprisingly, is given as 31 (instead of the expected 32), much to our amazement this is a correct answer as well, and therefore, **1, 2, 4, 8, 16, 31** is also a legitimate sequence.

We will justify this strange appearance later, but in the meantime, let us first find the succeeding numbers in this unexpected sequence. We set up a table of differences in Figure 52 showing the differences between terms of a sequence. We begin with the given sequence up to 31, and then work backward once a pattern is established (here, at the third difference).

| Original Sequence | 1 | | 2 | | 4 | | 8 | | 16 | | 31 |
|---|---|---|---|---|---|---|---|---|---|---|---|
| First Difference | | 1 | | 2 | | 4 | | 8 | | 15 | |
| Second Difference | | | 1 | | 2 | | 4 | | 7 | | |
| Third Difference | | | | 1 | | 2 | | 3 | | | |
| Fourth difference | | | | | **1** | | **1** | | | | |

**Figure 52**

With the fourth differences forming a sequence of constants, we can reverse the process (turn the table upside down as shown in Figure 53) and extend the third differences a few more steps with 4 and 5, to add after our number 31, the numbers 57 and 99.

| | | | | | | | | | | | | | | | |
|---|---|---|---|---|---|---|---|---|---|---|---|---|---|---|---|
| Fourth Difference | | | | | 1 | | 1 | | 1 | | 1 | | | | |
| Third Difference | | | | 1 | | 2 | | 3 | | **4** | | **5** | | | |
| Second Difference | | | 1 | | 2 | | 4 | | 7 | | **11** | | **16** | | |
| First Difference | | 1 | | 2 | | 4 | | 8 | | **15** | | **26** | | **42** | |
| Original Sequence | 1 | | 2 | | 4 | | 8 | | 16 | | **31** | | **57** | | **99** |

**Figure 53**

The bold-type numbers are those that were obtained by working backward from the third difference sequence. The general term is a fourth-power expression as we had to go to the third differences to get a constant. The general term ($n$) is: $\dfrac{n^4 - 6n^3 + 23n^2 - 18n + 24}{24}$.

One should not think that this sequence is independent of other parts of mathematics. Consider the Pascal triangle[27] shown in Figure 54.

1
1 1
1 2 1
1 3 3 1
1 4 6 4 1
1 **5 10 10 5 1**
1 6 15 20 15 6 1
1 7 21 35 35 21 7 1
1 8 28 56 70 56 28 8 1

**Figure 54**

Consider the horizontal sums of the rows of the Pascal triangle to the right of the bold line:

1, 2, 4, 8, 16, **31**, 57, 99, 163. The bold numbers provide a sum of 31.

Before we enter the geometric aspects of the number 31, we should take note that the sum of the first eight digits of the value of $\pi = 3 + 1 + 4 + 1 + 5 + 9 + 2 + 6 = 31$. Furthermore, the value of $\pi^3 = 31.006276680299820175476315067101$ is very close to the number 31.

A geometric interpretation should help convince you of the beauty and consistency inherent in mathematics. The number of regions into which a circle can be partitioned by joining points on the circle. Make sure that no three lines meet at one point, or else you will lose a region.

When we do this with six points as shown in Figure 55, we find that there are 31 regions. The chart in Figure 56 summarizes the number of regions based on specific points on the circle.

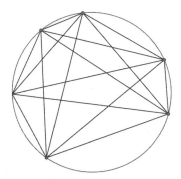

**Figure 55**

| Number of points on the circle | Number of regions into which the circle is partitioned |
|:---:|:---:|
| 1 | 1 |
| 2 | 2 |
| 3 | 4 |
| 4 | 8 |
| 5 | 16 |
| **6** | **31** |
| 7 | 57 |
| 8 | 99 |

**Figure 56**

We can take this further geometrically because the number 31 appears as the lowest prime centered pentagonal number as we see in Figure 57.

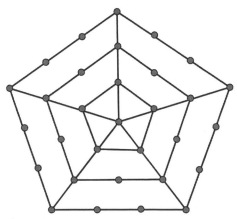

**Figure 57**

We know that there are certain polygons that can be constructed with an unmarked straightedge and compasses, such as an equilateral triangle, square, a regular pentagon, but not a regular heptagon (a seven-sided regular polygon). To date, we know that there are 31 regular polygons with an odd number of sides that can be constructed with an unmarked straightedge and compasses.

We complete our amazement with the number 31 by realizing that it is a *self-number*, which is a number that cannot be expressed as a number plus that number's digits. For example, 21 is not a self-number because it can be expressed as the number 15 plus the sum of 15's digits, that is, $15 + 1 + 5 = 21$.

## THE NUMBER 32

Those familiar with Fahrenheit are fully aware that 32° is the freezing point for water. We all know that a full set of teeth in the human body is 32. Ludwig van Beethoven wrote 32 piano sonatas between 1795 and 1822. The 32nd state to join the United States was Minnesota on May 11, 1858.

Mathematically, the number 32 can also be expressed as $32 = 1^1 + 2^2 + 3^3$, or $32 = 2^4 + 4^2$. Interestingly enough, the number 32 can also be expressed as $32 = 3^4 - 7^2$.

The number 32, which is $2^5$, is the highest power of 2 where all the digits are prime numbers. Using the digits of the number 32, we can establish two prime numbers, namely, $23 = 32 - 3^2$ and $41 = 32 + 3^2$.

The famous French mathematician, Pierre de Fermat (1607–1665), established a sequence of prime numbers commonly referred to as Fermat primes, and are symbolized as $F_n = 2^{2^n} + 1$, where $n \geq 0$. The Fermat primes are: 3, 5, 17, 257, and 65,537. The number 32 amazingly appears when we take the product of these Fermat primes, which is:

$$3 \times 5 \times 17 \times 257 \times 6,537 = 4,294,967,295 = 2^{32} - 1.$$

## THE NUMBER 33

In the days when phonograph records were still the prime source for music at home, the premier version was known as 33s, referring to the fact that those records rotated in the revolution speed of $33\frac{1}{3}$ revolutions per minute. In sports, the Indianapolis 500 typically has 33 racing competitors. The state of Oregon was the 33rd state to join the United States on February 14, 1859.

In mathematics, we recognize immediately that the number 33 is not a prime number, the fact that it has exactly two factors, namely, 3 and 11, qualify it to be a *semiprime* number and strangely enough is followed by two other semiprime numbers, namely, 34 and 35. We would have to wait until the number 85 until we can once again establish a triple of consecutive semiprime numbers, namely, 85, 86, and 87. The next such triples of consecutive semiprime numbers are (93, 94, 95); (141, 142, 143); (201, 202, 203); (213, 214, 215); (217, 218, 219); and so forth.

The number 33 also has a curious distinction in that it can be expressed as a sum of the factorials of the first four natural numbers, as: $33 = 1! + 2! + 3! + 4! = 1 + 2 + 6 + 24$.

The number 33 also carries some of the peculiarities that have been discovered by mathematicians, such as that the number 33 is the largest number where $2^{33} = 8,589,934,592$ and $5^{33} = 116,415,321,826,934,814,453,125$, each contains no zeros.

With the aid of a computer, we can find some astonishing relationship such as:
$$33 = 8,866,128,975,287,528^3 + (-8,778,405,442,862,239)^3 + (-2,736,111,468,807,040)^3.$$

There are some curious ways that the number 33 can generate prime numbers such as the following:

$33 \times 4^{33} + 1 = 2,434,970,217,729,660,813,313$

$33 \times 14^{33} + 1 = 2,33191,488,102,653,841,084,647,604,804,502,947,889,153$

$33 \times 34^{33} + 1 = 11,410,855,408,151,892,011,078,343,133,624,756,861,835,\\061,265,170,433$

But $33 \times \mathbf{24^{33}} + 1 = 116,274,616,654,336,630,549,169,399,022,016,360,650,\\005,512,193$ is *not* a prime number.

It is believed that there are no longer sequences of consecutive numbers amongst the natural numbers, which includes *no* prime numbers than the sequence of 33 numbers shown here:

1,328, 1,329, 1,330, 1,331, 1,332, 1,333, 1,334, 1,335, 1,336, 1,337, 1,338, 1,339, 1,340, 1,341, 1,342, 1,343, 1,344, 1,345, 1,346, 1,347, 1,348, 1,349, 1,350, 1,351, 1,352, 1,353, 1,354, 1,355, 1,356, 1,357, 1,358, 1,359, 1,360.

## THE NUMBER 34

In the history of the United States on January 29, 1861, Kansas was the 34th state to join the United States. One of the more famous streets in New York City is 34th Street, which is the home to the largest department store, Macy's.

The number 34 enjoys a few special distinctions in mathematics, one of which is that it is the ninth member of the Fibonacci sequence: 1, 1, 2, 3, 5, 8, 13, 21, **34**, ..., and as introduced previously, the number 34 is the second member of a semiprime triplet: 33, 34, 35.

The number 34 is the smallest number that can be expressed as a sum of two primes in four different ways, as follows: $34 = 3 + 31 = 5 + 29 = 11 + 23 = 17 + 17$. Furthermore, the number 34 is the smallest number that can be expressed as a sum of distinct primes, each of which is comprised only of odd digits, as follows: $34 = 3 + 5 + 7 + 19$. (Notice, all the single-digit odd numbers have been used exactly once.) The number 34 is also equal to the sum of the squares of two consecutive prime numbers: $34 = 3^2 + 5^2$, which, by the way, are Fibonacci numbers.

In our discussion of the number 16 we introduced the magic square with 16 cells (Figure 58) that was attributed to the German artist Albrecht Dürer, where the number 34 dominates, as it is the sum of each row, column, and diagonal.

We recall that a regular heptagon is one with seven equal length sides and angles. The number 34 is shown in Figure 59 as a heptagonal number.

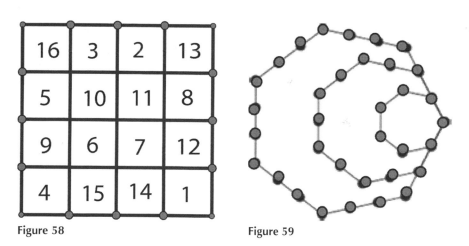

| 16 | 3 | 2 | 13 |
|----|----|----|----|
| 5 | 10 | 11 | 8 |
| 9 | 6 | 7 | 12 |
| 4 | 15 | 14 | 1 |

**Figure 58**                    **Figure 59**

## THE NUMBER 35

The number 35 has long had fame in the film industry because of the very popular 35mm film used in photography and motion pictures. The 35th state the join the United States was West Virginia on June 20, 1863.

We can use the first three primes represent the number 35 as follows: $35 = 2^3 + 3^3$. This can be taken a step further in that the number 35 can be written as a sum of odd primes in 35 different ways, such as, for example, here are three ways,

$$35 = 29 + 3 + 3 = 23 + 3 + 3 + 3 + 3 = 11 + 2 + 2 + 2 + 2 + 3 + 3 + 5 + 5.$$

The motivated reader may wish to find the remaining 32 ways to express the number 35 as the sum of odd primes numbers.

A cute relationship that can be constructed by the sum and the product of the digits of 35, each of which produces a number, which is 1 less than a perfect square. That is, $3 + 5 = 8 = 3^2 - 1 = 8$, and $3 \times 5 = 5^2 - 1 = 15$.

The number 35 in a somewhat strange way can lead us to the next number, 36. The prime factors of 35 are 5 and 7. These prime factors are used as follows: $5^4 \times 7$ produces 4,375; while the prime factors of 36 are 2 and 3 (as $2^2 \times 3^2 = 36$) and are the same prime factors used as follows: $3^7 \times 2$ produces 4,374. In other words, using prime factors the number 35 and its successor 36 were able to generate two consecutive numbers 4,374 and 4,375. So onward to the number 36.

## THE NUMBER 36

The number 36 is well known as the number of inches in one yard. From a historical perspective on October 31, 1864, Nevada was the 36th state to join the United States. On a standard roulette wheel there are 36 numbers excluding the zeros. A standard beer barrel in the Great Britain has a volume of 36 gallons. We need to wonder if Superman knew that the atomic number of krypton is 36.

Mathematically, the number 36 is a square number, since $6^2 = 36$, and also the smallest square number that can be expressed as a sum of a pair of twin prime numbers, namely $36 = 17 + 19$. Furthermore, beyond the sum of twin primes, the number 36 is the smallest number that can be expressed as a sum of two distinct primes in three different ways, as follows: $36 = 5 + 31 = 7 + 29 = 13 + 23$. We could also look at this in an extended fashion by noting that the number 36 can be expressed in two different ways as a sum of consecutive prime numbers: $36 = 5 + 7 + 11 + 13$, or $36 = 17 + 19$.

Partnering with its reversal, the number 36 is the smallest multidigit number that generates prime numbers in the following fashion: $36 \times 63 - 1 = 2,267$, and $36 \times 63 + 1 = 2,269$, producing a pair of twin primes, that is, the 336th prime and the 337th prime, respectively.

The number 36 is also the eighth triangular number, as we can see in Figure 60. This of course can be shown arithmetically as follows: $1 + 2 + 3 + 4 + 5 + 6 + 7 + 8 = 36$. Another way of saying this is that the number 36 could also be considered equal to the sum of the first four odd numbers and the first four even numbers.

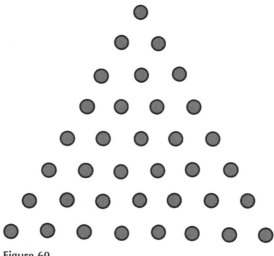

**Figure 60**

The number 36 also has the distinction of being the largest two-digit number that is divisible by both the product of its digits ($3 \times 6$) and the sum of its digits ($3 + 6$). Somewhat analogously, we can show a neat product and a sum that equals the number 36:

$$1^2 \times 2^2 \times 3^2 = 1 \times 4 \times 9 = 36, \text{ and } 1^3 + 2^3 + 3^3 = 1 + 8 + 27 = 36.$$

In probability, when we toss a pair of dice, the number of possible outcomes is $6 \times 6 = 36$. The number 36 also is a distinct place in geometry since it is the measure of each exterior angle regular pentagon.

On a less happy note, if we take the sum the numbers from 1 to 36, we get a total of 666, which is "the number of the beast" (see page 170).

## THE NUMBER 37

It is believed that William Shakespeare had written 37 plays. Shakespeare mentions the word *mathematics* only once in all his writings and that appears in "The Taming of the Shrew" on line 37 of Act I, Scene 1 in the sentence:

Balk logic with acquaintance that you have,
And practice rhetoric in your common talk;
Music and poesy use to quicken you;
The **mathematics** and the metaphysics—(Line 37)
Fall to them as you find your stomach serves you.

In medicine, 37°C (98.6° Fahrenheit) is considered the mean healthy human body temperature. This was first established by the German physician Carl

Reinhold August Wunderlich (1815–1877). On March 1, 1867, Nebraska was the 37th state to join the United States.

Here is another peculiarity that further demonstrates the uniqueness of the number 37. In a nonleap year, August is the only month where the sum of the letters in the name and the number of days in the month is a prime number, the number 37.

The number 37 is the smallest prime number formed by two different odd prime digits. (Remember, the number 1 is not a prime number.)

When we omit the number 8 the initial sequence of numbers is: $1 + 2 + 3 + 4 + 5 + 6 + 7 + 9 = 37$. Once again, omitting the number 8 we get the curiosity $3 \times 12,345,679 = 37,037,037$, which is not a prime number because $37,037,037 = 3 \times 37 \times 333,667$.

We should initially note that the number 37 is the only prime number, whose reciprocal has a period of length 3, as we see as follows: $$\frac{1}{37} = 0.027\ 027\ 027\ 027\ 027....$$

Not only is the number 37 a prime number, however, its reversal, 73, is also a prime number. Furthermore, each of these prime numbers can be expressed as a sum of 2 squares: $37 = 1^2 + 6^2$, and $73 = 3^2 + 8^2$.

Curiously, the sum of the prime divisors of the two numbers following the number 37, namely 38 and 39 have a sum of 37, as we see here: $38 = 2 \times 19$, and $39 = 3 \times 13$, so that $2 + 19 + 3 + 13 = 37$.

The number 37 is the largest prime factor of any six-digit repdigit number. For example, consider the repdigit number $888,888 = 37 \times 24,024$; or as another example, $333,333 = 37 \times 9009$.

Notice how using the sequence (considering the bases and the exponents) of prime numbers up to the number 37 can be used to generate another prime: $2^3 + 5^7 + 11^{13} + 17^{19} + 23^{29} + 31^{37} = 15,148,954,872,646,850,196,557,152,$ $427,604,893,685,308,877,022,260,348,791$, which is a very large prime number.

The number 37 lends itself to a rather amazing property. If you take any multiple of 37 and break it up into two numbers and then find the sum of the cubes each of these two parts, the result will always be a multiple of 37. Let's consider a few examples: $37 \times 5 = 185$. Suppose we split the number into 18 and 5 and then cubes each part: $18^3 + 5^3 = 5957 = 37 \times 161$. This time take the number $37 \times 8 = 296$ and split the number as 2 and 96. The sum of the cubes is: $2^3 + 96^3 = 8 + 884,736 = 884,744 = 37 \times 23,912$, once again a multiple of 37. An ambitious reader might want to try this with other multiples of 37.

The number 37 has some peculiar characteristics. For example, suppose we take a three-digit number that is divisible by 37, such as $296 = 8 \times 37$. We will find the numbers 962 and 629, which are two numbers where the digits have been cyclically permutated (same order, but different starting point) from the original number 296. Each of these two numbers is also divisible by 37 because $962 = 26 \times 37$, and $629 = 17 \times 37$.

The number 37 can also be found by taking the sum of the first five consecutive composite numbers: $37 = 4 + 6 + 8 + 9 + 10$. We can also create 37 as a sum using only the number 3 as follows: $37 = 33 + 3 + \frac{3}{3}$.

The number $2^{37} = 137,438,953,472$ is not a prime number; however, the reverse of this number, $274,359,834,731$, is a prime number.

Another curious characteristic of the number 37 is found by the following: $37^2 + 1 = 1,370$, whose largest prime factor is 137. Also, $2 \times 37 = 74$, which is one greater than the reverse of 37, namely, 73.

As we mentioned earlier on several other number explorations, the French mathematician Sophie Germain (1776–1831) established a special kind of prime number that today carries her name as a *Sophie Germain prime*, which is a prime number, *n*, where the number $2n + 1$ is also a prime number. The first few Sophie Germain primes are: 2, 3, 5, 11, 23, 29, 41, 53, 83, 89, 113, and so on. A reason for bringing it up at this point is that there are 37 Sophie Germain primes less than 1,000.

We close our discussion of the number 37 with the following pattern:

$$37\times(1+1+1) = 111$$

$$37\times(2+2+2) = 222$$

$$37\times(3+3+3) = 333$$

$$37\times(4+4=4) = 444$$

$$37\times(5+5+5) = 555$$

$$37\times(6+6+6) = 666$$

$$37\times(7+7+7) = 777$$

$$37\times(8+8+8) = 888$$

$$37\times(9+9+9) = 999$$

**Equation 6**

## THE NUMBER 38

The number 38 in history at has a very curious position. It is the 38th parallel north, which separates North Korea and South Korea. In the history of the United States on August 1, 1876, the 38th state to join the union was Colorado.

We know that in mathematical terms, the number 38 is not a prime number. Yet we can show that 38 is related to prime numbers in that the sum of the squares of the first three prime numbers, namely, $2^2 + 3^2 + 5^2 = 38$. Furthermore, we can also create a prime number from the digits of 38 as follows: $3^2 + 8^{19} = 9 + 144,115,188,075,855,872 = 144,115,188,075,855,881$, which is a prime number.

When paired with its predecessor, the numbers 37 and 38 form the first pair of numbers, where each is not divisible by the sum of its digits.

It is conjectured that the number 38 is the largest even number that can be expressed in only one way as the sum of two distinct prime numbers, as $38 = 31 + 7$. It is also the largest even number that can be expressed as the sum of two odd numbers, where exactly at least one of the numbers is a prime number. An ambitious reader may want to check and verify that 38 is the largest even number that is not the sum of two odd composite numbers.

The number 38 is the average of the first prime numbers up to 83.

The number 38 is a composite number, and if we add 1s onto this number we will always have another composite number such as 381, 3,811, 38,111, 381,111, 3,811,111.

## THE NUMBER 39

In the history of the United States in 1787, the Constitution was signed by 39 of the 55 members at the Philadelphia convention. On November 2, 1889, the 39th state to join the union was North Dakota.

From a mathematical perspective the number 39 can be expressed as the sum of the product and the sum of its digits as we show with the following $(3 \times 9) + (3 + 9) = 39$.

The number 39 can also be expressed as a sum of consecutive prime numbers: $3 + 5 + 7 + 11 + 13 = 39$, and also as a product of the first and last of those prime numbers: $3 \times 13 = 39$.

The number 39 can also be expressed as the sum of the first 3 powers of 3 as we can see as follows: $3^1 + 3^2 + 3^3 = 39$.

## THE NUMBER 40

The number *Forty* is the only number whose word in the English language has the letters appear in alphabetical order. The temperature −40° is the same in Fahrenheit and Celsius. The English word *quarantine* comes from the Middle Ages when European ships would not let passengers off for 40 days during the bubonic plague. The word for 40 in Italian is quaranta, hence, the name quarantine. According to the Bible, the number 40 seems to have had some unusual significance. For example, it rained for 40 days and 40 nights causing the flood, where Noah mounted his ark to save himself and his selected animals. It is said that the Jewish people wandered the desert for 40 years. Furthermore, Jesus is said to have fasted in the wilderness for 40 days. It should be noted that on November 2, 1889, South Dakota was the 40th state join the United States.

Seen mathematically, the number 40 can be represented as the sum of four pentagonal numbers, namely $1 + 5 + 12 + 22 = 40$. The number 40 can also be expressed using the first three prime numbers in order, as we see with $2^3 \times 5 = 40$.

The number 40 can also be represented as progressive powers of base 3, as follows: $3^0 + 3^1 + 3^2 + 3^3 = 1 + 3 + 9 + 27 = 40$.

Geometrically, we can see that the number 40 is an octagonal number as shown in Figure 61.

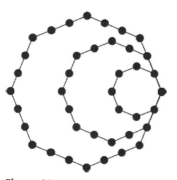

**Figure 61**

## THE NUMBER 41

Wolfgang Amadeus Mozart's last symphony was (*The Jupiter*) Symphony No. 41; it was also his longest symphony. If we look hard enough, we can even find the number 41 appearing in sports. For example, in Super Bowl XLI it was the first time that the final score of both teams were prime numbers: Indianapolis Colts

29, and the Chicago Bears 17. The sum of the common American coins: penny, nickel, dime, and quarter, have a sum of 41 cents. The number 41 also highlights Montana as the 41st state to join the United States on November 8, 1889.

We can also give the number 41 a special place in our number system by saying that it is the smallest prime whose first digit is a composite number, namely, 4. Number 41 is also the largest two–digit prime number, where if you add 2 to each digit separately, prime numbers will result, namely, 43 and 61 are prime numbers.

The number 41 is also the smallest prime number the sum of whose digits, 5, and the difference of the digits, 3, form a pair of twin primes, which we recall, are two prime numbers whose difference is 2.

The number 41 can be expressed as the sum of the first six prime numbers, as $2 + 3 + 5 + 11 + 13 = 41$, and also can be expressed, simply, as a sum of three other prime numbers, as $11 + 13 + 17 = 41$.

Furthermore, the number 41 added to the factorial of its reversal is a prime number: $41 + 14! = 41 + 87,178,291,200 = 87,178,291,241$, which is a prime number.

A curiosity about the number 41 was discovered, where it was shown equal to the sum of the first three factorials plus the first three self-powers: $1! + 2! + 3! + 1^1 + 2^2 + 3^3 = 1 + 2 + 6 + 1 + 4 + 27 = 41$. Also, the sum of the factorials of the digits is equal to a perfect square, as $4! + 1! = 25 = 5^2$.

Merely as a curiosity, we find that $41 \times 2,439 = 99,999$.

Number 41 can also be expressed as the sum of two consecutive squares, such as $4^2 + 5^2 = 41$.

The number 41 is the smallest prime number whose cube can be expressed as a sum of three cubes in two different ways, as we show here: $41^3 = 40^3 + 17^3 + 2^3 = 33^3 + 32^3 + 6^3$. We should also note that 41 is the smallest prime number that cannot be expressed as the difference between two powers of 2 and 3. As a counterexample, the prime number $19 = 3^3 - 2^3$.

The number 41 is the smallest integer whose reciprocal can produce repeating digits of length 5, which is: $\dfrac{1}{41} = 0.02439\,02439\,02439\,02439\,02439\,02439\ldots$.

The number of noncomposite numbers up to and including 41 is the reversal of 41, namely 14. Another 41 curiosity is that $2^{29} = 536,870,912$, whose digits sum is 41.

The famous Swiss mathematician Leonhard Euler (1707–1783) developed the formula $x^2 + x + 41$, which generates primes beginning with the number 41, when $x = 0$. This continues on consecutively until $x = 39$ when the prime number 1,601 is reached. When $x = 40$, the formula results in $1681 = 41^2$, which is certainly not a prime number.

This can be somewhat generalized, if we consider the formula $x^2 + x + n$, where aside from $n = 41$, we select any of the following numbers for the value of $n$: 2, 3, 5, 11, and 17, prime numbers will be generated by number up to

$n - 2$. For example, suppose we let $n = 5$, then prime number will be obtained by $x^2 + x + 5$ for values of $x$ up to $5 - 2 = 3$.

For $x = 0$, we get prime number 5.
For $x = 1$, we get prime number 7.
For $x = 2$, we get prime number 11.
For $x = 3$, we get prime number 17.
However, for $x = 4$, we get 25, which is not a prime number.

A nice little technique can be created beginning with the number 41. Begin with the number 41, then add 2 to it to get $41 + 2 = $ **43**, then add 4 to that number to get $43 + 4 = $ **47**, then add 6 to get **53**, then add 8 to get **61**, then add 10 to get **71**, then add 12 to get **83**; the next number using this scheme would be **97**. By continuing this process 40 times you will have a string of 40 prime numbers. When you do it for the 41st time you will end up with a composite number.

It is believed that highest power of 11 that contains no zeros is $11^{41} = 4,978,518,112,499,354,698,647,829,163,838,661,251,242,411$.

Here is a peculiar challenge that appears to be true, but we leave it to the motivated reader to test out several possibilities. Select a pair of numbers whose sum is 41. Square one of the numbers and add the other one to the squared number, and you should end up with a prime number. For example, suppose we select the numbers 33 and 8, whose sum is 41. Following this procedure: $33^2 + 8 = 1,097$, which is a prime number. Or we can do it in the other direction: $8^2 + 33 = 97$, which is also a prime number.

A quick method to multiply by 41 is as follows: Quadruple the number you wish to multiply by 41, and then multiply the result by 10 and add the original number. To see how this works consider the following example: to multiply 47 by 41, quadruple 47, which yields 188, then multiply by 10 to get 1880, and then add the original number 47 to get 1927, which is equal to $47 \times 41$.

## THE NUMBER 42

One of the most famous baseball players who has made history, is the Brooklyn Dodger's Jackie Robinson, whose uniform number was 42, a number that is now retired from use by any future player. One of the most famous streets in New York City is 42nd Street. Further to our discussion of the number 42, we can mention that in the Japanese culture the number 42 is considered an unlucky number—largely because when the numerals (4, 2) are pronounced separately it is *shi ni*, which sounds like "dying." On a more positive note, the 42nd state to join the United States was the state of Washington on November 11, 1889.

The number 42 is a composite number and yet the product of three distinct prime numbers, namely, $42 = 2 \times 3 \times 7$. The number 42 is also the sum of the first six positive even integers, as $42 = 2 + 4 + 6 + 8 + 10 + 12$. And further, the number 42 is also equal to the sum of its nonprime proper divisors: $42 = 1 + 6 + 14 + 21$. In 2019, mathematicians found that the number 42 can be expressed as the sum of three cubes as follows: $(-80,538,738,812,075,974)^3 + (80,435,758,145,817,515)^3 + (12,602,123,297,335,631)^3$.

The number 42 also lends itself to a curious relationship between the cubes of two prime twins, namely, $42^2 = 11^3 + 433 = 13^3 - 433$.

We can also create an interesting group of five prime numbers using the number 42:

> $1 + 42 = 43$, which is the 14th prime number
> $1 + 4242 = 4243$, which is the 582nd prime number
> $1 + 424242 = 424243$, which is the 35,721st prime number
> $1 + 42424242 = 42424243$, which is the 2,571,845th prime number
> $1 + 4242424242 = 4242424243$, which is also a prime number!

## THE NUMBER 43

In England, the year 43 CE is of significance, as it is the year that the Romans invaded Great Britain, and seven years later founded London. Another everyday experience attached to the number 43 occurred at the beginning of the introduction to telephone area codes, where two adjacent states had area codes that were multiples of 43: Pennsylvania's area code was 215 (which is $5 \times 43 = 215$) and Maryland's area code was the real one (which is $7 \times 43 = 301$). On July 3, 1890, the 43rd state to join the United States was Idaho. The telephone code for Austria is +43.

Not only is the number 43 a prime number, but it is also the second partner of the pair of twin primes 41 and 43.

The number 43 can be used to form a prime number as $4^3 + 43 = 107$, which is also a prime number.

Strangely, $43! = $ **60,415,263,063,373,835,637,355,132,132,068,513,997,507,264,512**,000,000,000, where the number formed by the first 43 digits is a prime number—as shown in bold. To take this a step further, if we take the number 43 and raise it to the $(4 + 3)$ power the result is a digit sum of 43, as we can see with the following: $43^7 = 271,818,611,107$, where $2 + 7 + 1 + 8 + 1 + 8 + 6 + 1 + 1 + 1 + 0 + 7 = 43$. It has also been found that a group of five primes that "fit" with the number 43 are as follows:

# 43
# 4483
# 444883
# 44448883
# 4444488883

**Equation 7**

Mathematicians often discover personal curiosities with numbers. For example, the English mathematician Augustus De Morgan (1806–1871) was a man of wisecracks, which he felt were humorous. There were times when he found mathematical coincidences such as with the number 43. He claimed to have the good fortune of being $x$ years old in the year $x^2$. That is, he was 43 years old in 1849, and $43^2 = 1,849$. Some mathematicians have also found that 43 is the smallest prime number that is not the sum of two palindromic numbers. For example, suppose we consider the next smaller prime number which is 41 and can be expressed as the sum of two palindromic numbers, namely $33 + 8 = 41$.

Another such guarantee is if we see the number 43 as the smallest non-palindromic prime that when its reversal is subtracted from it results in a square number, as we can see with $43 - 34 = 9 = 3^2$. Another peculiarity to consider is that the number 43 is the only prime number, where using its digits sequentially as $4^4 - 3^3 = 229$, which is a prime number!

Geometrically, the number 43 presents itself as a centered heptagonal number as shown in Figure 62, where the 43 dots are exhibited in heptagonal form.

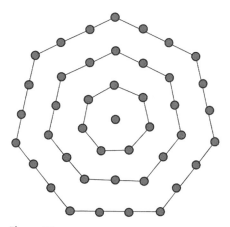

**Figure 62**

## THE NUMBER 44

On July 10, 1890, the state of Wyoming was the 44th state to join the United States. The international telephone code for the United Kingdom is +44. There are 44 candles in a complete box of Hanukkah candles.

The number 44 is a palindromic number as well as an octahedral number, which is the number of closely packed spheres that fit into a regular octahedron. Octahedral numbers can be found with the formula: $\dfrac{n(2n^2 + 1)}{3}$. In this case $n = 4$.

Furthermore, the first 44 numbers can be partitioned into four sets so that any two members of each set form a unique sum within that set. These sets are: $\{1,3,5,15,17,19,26,28,40,42,44\}$, $\{2,7,8,18,21,24,27,33,37,38,43\}$, $\{4,6,13,20,22,23,25,30,32,39,41\}$, and $\{9,10,11,12,14,16,29,31,34,35,36\}$.

We can also generate a prime number with the number 44 as follows: $44^{4 \times 4} + 1 = 197,352,587,024,076,973,231,046,657$.

## THE NUMBER 45

The 45th state to join the United States was Utah on January 4, 1896.

The number 45 is a triangular number, which produces the following sum: $0 + 1 + 2 + 3 + 4 + 5 + 6 + 7 + 8 + 9 = 45$. This triangular number can also be written as a sum of 2 squares as follows: $6^2 + 3^2 = 36 + 9 = 45$.

Quite unexpectedly, we can subtract the first five powers of 2 from the number 45 and the results will be five prime numbers as we show here: $45 - 2 = 43$, $45 - 2^2 = 41$, $45 - 2^3 = 37$, $45 - 2^4 = 29$, $45 - 2^5 = 13$.

The number of 45 is a hexagonal number as shown in Figure 63. The formula that can be used to determine hexagonal numbers is $n(2n - 1)$. When $n = 5$, we get the hexagonal number 45. The first few hexagonal numbers are 1, 6, 15, 28, and 45.... The relationship between hexagonal numbers $H_n$ and triangular numbers $T_n$ is $H_n = 4T_{n-1} + n$.

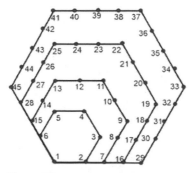

**Figure 63**

## THE NUMBER 46

The 46th state to join United States was Oklahoma on November 16, 1907.

The number 46 does not lend itself to any special mathematical aspects within the scope of this book.

## THE NUMBER 47

The 47th state to join the United States was New Mexico on January 27, 1912. In the music world, a concert harp has 47 strings.

The number 47 lends itself to a curious peculiarity, that is $47 + 2 = 49$, and $47 \times 2 = 94$, which are reversals of one another.

The number 47 lends itself to some oddities, such as: $4^3 + 7^3 = 407$.

In the list of prime numbers, the first 0 digit is reached after counting 47 digits.

The number 47 is equal to the sum of the primes: $2 + 3 + 5 + 7 + 11 + 19$.

## THE NUMBER 48

The state of Arizona was the 48th state to join the United States on February 14, 1912. There are 48 minutes of playing time in a National Basketball Association game. The *Well-Tempered Clavier*, by Johan Sebastian Bach, is often referred to as "The 48th" because it consists of a prelude and a fugue in each minor and major key for a total of 48 pieces.

The number 48 is the smallest number that has exactly 10 divisors, which are: 1, 2, 3, 4, 6, 8, 12, 16, 24, and 48.

The composite number 48 is an even number that can be expressed as a sum of two prime numbers in five different ways, as we show here: $5 + 43 = 48, 7 + 41 = 48, 11 + 37 = 48, \quad 17 + 31 = 48, 19 + 29 = 48,$ The number 48 is the smallest number that can generate two prime numbers as follows:

$7^{48} + 48 = 36,703,368,217,294,125,441,230,211,032,033,660,188,849,$
is a prime number, and
$7^{48} - 48 = 36,703,368,217,294,125,441,230,211,032,033,660,188,753,$
is also a prime number.

The number 48 is a sum of four distinct primes, namely, 5, 7, 17 and 19, where the sum of any three of these four prime numbers yields a prime number, such as, for example, $5 + 7 + 17 = 29$, or $5 + 17 + 19 = 43$, and so on.

## THE NUMBER 49

The 49th state to join the United States was Alaska on January 3, 1959. It should be noted that it is the 49th parallel that largely separates the United States from Canada. Yet, for many people the San Francisco 49ers football team would come to mind, which got its name from the California Gold Rush of 1849. Sports fans will also remember that the number 49 had a great deal of significance in professional boxing because Rocky Marciano was the only heavyweight champion who never lost a bout after 49 wins.

The number 49 is the smallest square that can be expressed as the sum of three consecutive primes as we see with $13 + 17 + 19 = 49$.

The number 49 can also be considered the smallest prime number squared that is a sum of consecutive composite numbers in three various ways, as shown: $15 + 16 + 18 = 49, 25 + 24 = 49$, and $4 + 6 + 8 + 9 + 10 + 12 = 49$.

From a mathematical perspective, we know that 49 is the square of 7. Yet the relationship of 49 to its partner square, 7, is that $49^2 = 2,401$, where the sum of its digits, $2 + 4 + 0 + 1 = 7$. It is also to be noted that the number 49's digits are each perfect squares. Curiously, $49^3 = 117,649$. Notice the last two digits are 49.

The reciprocal of 49, that is,

$$\frac{1}{49} = 0.\overline{020408163265306122448979591836734693877551}$$

consists of 42 digits which repeat indefinitely. When we look at the first 10 places, we notice the powers of 2 as 02 04 08 16 32; it then continues with overlaps, which are not so visible unless you can carry over digits as we can see from this addition:

$$
\begin{array}{r}
02 \\
04 \\
08 \\
16 \\
32 \\
64 \\
128 \\
256 \\
512 \\
1024 \\
+ \qquad 2048 \\
\end{array}
$$

**Equation 8**

An interesting feature of this 42 repeating digit-span appears when we multiply it by 2, 3, 4, and so forth.

$$2 \times 020408163265306122448979591836734693877551 =$$
$$408163265306122448979591836734693877551 02$$
$$3 \times 020408163265306122448979591836734693877551 =$$
$$6122448979591836734693877551 0204081632653$$
$$4 \times 020408163265306122448979591836734693877551 =$$
$$8163265306122448979591836734693877551 020$$
$$5 \times 020408163265306122448979591836734693877551 =$$
$$10204081632653061224489795918367346938 77755$$

Notice how each multiple of the repeating decimal provides a cyclic repeat of the original number, yet starting at a different digit.

## THE NUMBER 50

The 50th state to join the United States on August 21, 1959, was Hawaii, and then, of course, the current United States of America has 50 states. The flag of the United States has 50 stars. When a Jubilee is celebrated, it typically indicates a 50-year celebration. A golden anniversary celebrates the 50th year of marriage. The Christian feast of Pentecost takes place on the 50th day after the Easter season.

The number 50 can be expressed as a sum of two squares in two different ways, namely, $1^2 + 7^2 = 50$ and $5^2 + 5^2 = 50$. The number 50 can also be expressed as the sum of three squares, as $3^2 + 4^2 + 5^2 = 50$. And even more, the number 50 can also be expresses as the sum of four squared numbers, as $1^2 + 2^2 + 3^2 + 6^2 = 50$.

Using all the initial primes exactly once, the number 50 can be expressed as the sum of these in two different ways: $2 + 5 + 7 + 17 + 19 = 50 = 3 + 11 + 13 + 23$.

## THE NUMBER 51

In American history one of the significant contributions are the Federalist Papers to which Alexander Hamilton contributed 51 essays.

Although this would not be easily created, a regular polygon with 51 sides can be constructed with an unmarked straightedge and pair of compasses.

The number 51 is also considered a pentagonal number, which we can see in Figure 64, where we show 51 dots forming a pentagon.

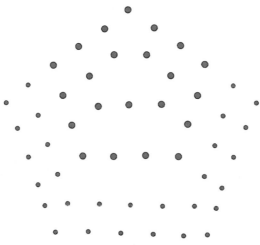

**Figure 64**

## THE NUMBER 52

The number 52 typically refers to the number weeks in the year, or to the number of cards in a standard pack of playing cards. A standard piano has 52 white keys.

The number 52 does not have many peculiarities in mathematics, although we can use it to generate a prime number in a rather interesting way: $52! + 52 + 1 = 80,658,175,170,943,878,571,660,636,856,403,766,975,$ $289,505,440,883,277,824,000,000,000,053$, which is a prime number.

## THE NUMBER 53

In the American National Football League, no team can have more than 53 players on the roster.

The number 53 is the 16th prime number, and of the first 53 prime numbers their sum is 5830, which is divisible by 53, that is, $\dfrac{5830}{53} = 110$.

The sum of the digits of $53^3$ is the reverse of 53, namely 35, since $53^3 = 148877$, and $1 + 4 + 8 + 8 + 7 + 7 = 35$.

When we consider the prime factors of $1,111,111,111,111$, they are 53, 79, and $265,371,653$, of which the number 53 is the smallest.

Number 53 cannot be the first two digits of a three-digit number which is prime; in other words the number $53n$ cannot be a prime number.

When the number 53 is the power to which the number 2 is taken as in $2^{53}$ the result is 9,<u>00</u>7,199,254,740,992, which is the smallest power of 2 that contains two consecutive zeros.

When we raise the number 53 to the 7th power, we find that $53^7 = 1,174,711,139,837$, and then the amazing result is that the sum of the digits of this number is $1 + 1 + 7 + 4 + 7 + 1 + 1 + 1 + 3 + 9 + 8 + 3 + 7 = 53$.

The number 53 is the only two-digit prime number, where the sum of its digits $5 + 3 = 8 = 2^3$ and the number 2 is the difference 53's digits, $5 - 3 = 2$.

The number 53 can be seen as the average of three consecutive double-digit primes:

$$\frac{47 + 53 + 59}{3} = 53,$$ and also, the average of seven consecutive primes:

$$\frac{41 + 43 + 47 + 53 + 59 + 61 + 67}{7} = 53.$$

Curiously, there are no twin primes (two consecutive primes that differ by 2) between $53^2$ and $54^2$, that is, between: 2,809 and 2,916.

## THE NUMBER 54

There is not much that can be said about the number 54 except that there are currently 54 sovereign countries on the African continent.

The number 54 can be written as a sum of three squares in three different ways:

$$1^2 + 2^2 + 7^2 = 54, \quad 3^2 + 3^2 + 6^2 = 54, \quad 2^2 + 5^2 + 5^2 = 54.$$

Sin $54° = 0.809016994375...$, and at first sight seems rather unimportant; however, when we double it, 2 Sin $54° = 1.6180339887...$, which just happens to be the Golden Ratio, that is $\frac{1 + \sqrt{5}}{2} \approx 1.6180339887....$ This is truly an unexpected relationship!

## THE NUMBER 55

The number 55 is the largest Fibonacci number that is also a triangular number. Recall the triangular numbers to see the number 55's position:

0, 1, 3, 6, 10, 15, 21, 28, 36, 45, **55**, <u>66</u>, 78, 91, 105, 120, 136, 153, 171, 190, 210, 231, 253, 276, 300, 325, 351, 378, 406, 435, 465, 496, 528, 561, 595, 630, <u>666</u>....

Here you can see the only three triangular numbers that are composed only of a multiple use of the same digit: 55, 66, and 666.

Here is a cute "trick" with 55: $5^3 + 5^3 = 250$, and $2^3 + 5^3 + 0^3 = 133$, and then $1^3 + 3^3 + 3^3 = 55$.

The number 55 allows a very unusual list of products as we can see in the following text:

| | |
|---|---|
| 55×91 = 5005 | 55×101 = 5555 |
| 55×93 = 5115 | 55×103 = 5665 |
| 55×95 = 5225 | 55×105 = 5775 |
| 55×97 = 5335 | 55×107 = 5885 |
| 55×99 = 5445 | 55×109 = 5995 |

**Equation 9**

## THE NUMBER 56

In the history of the United States, the Declaration of Independence was signed by 56 men. Baseball fans are surely aware of the 56-game hitting streak which the New York Yankees' Joe DiMaggio accomplished in 1941, from May 15 to July 16.

The number 56 is equal to the first six triangular numbers:
$1 + 3 + 6 + 10 + 15 + 21 = 56$.
The number 56 is also the sum of the six consecutive primes:
$3 + 5 + 7 + 11 + 13 + 17 = 56$.
The number 56 is double a perfect number, namely, $28 \times 2 = 56$.

## THE NUMBER 57

In mathematics the number 57 can be most appreciated at much higher levels than we present in this book. However, it is used in everyday life for the logo on products produced by the H. J. Heinz Company when they say: "Heinz 57 Varieties," and in New York City 57th St. is a famous street as it also houses Carnegie Hall. The number +57 is also the telephone code for the country of Colombia.

## THE NUMBER 58

The number 58 represents the sum of the first seven prime numbers: $2 + 3 + 5 + 7 + 11 + 13 + 17 = 58$. There are also 58 counties in the state of California. The telephone code for the country of Venezuela is +58.

## THE NUMBER 59

The oldest major league baseball player in the United States was Satchel Paige who played until age 59. In New York City the Queensboro Bridge is often referred to as the "59th Street Bridge."

The number 59 is the 17th prime number.

The number of value $15! + 1$ is divisible by 59, as shown here: $$\frac{1,307,674,368,001}{59} = 22,163,972,339$$

The number 59 can be expressed as a sum of three consecutive primes, namely, $17 + 19 + 23 = 59$.

The reversal of $59^{12} = 1,779,197,418,239,532,716,881$, which is $1,886,172,359,328,147,919,771$ is a prime number, and the sum of its digits $1 + 8 + 8 + 6 + 1 + 7 + 2 + 3 + 5 + 9 + 3 + 2 + 8 + 1 + 4 + 7 + 9 + 1 + 9 + 9 + 7 + 7 + 1 = 109$ is also a prime number.

The number 59 is the smallest number when divided by 2, 3, 4, 5, and 6 results with remainders of 1, 2, 3, 4, and 5, respectively.

The smallest power of 2 that results in a number where no digit appears only once is $2^{59} = 576,460,752,303,423,488$.

Here is a cute play with the number 59: we have $59^2 = 3,481$, yet written a different way we find that $3^4 = 81$.

## THE NUMBER 60

The number 60 is very dear to us because there are 60 minutes in an hour and 60 seconds in a minute. This could be related to the fact that $6 \times 60° = 360°$ is a complete revolution or circle, which probably stems from the fact that the Babylonians used the sexagesimal system (base 60) for their mathematical and astronomical work.

The number 60 could be considered quite "wealthy" because it has 12 divisors, namely, 1, 2, 3, 4, 5, 6, 10, 12, 15, 20, 30, and 60.

The number 60 is also the smallest number that can be expressed as the sum of two odd primes in six different ways: $7 + 53 = 60$, $13 + 47 = 60$, $17 + 43 = 60$, $19 + 41 = 60$, $23 + 37 = 60$, and $29 + 31 = 60$.

The number 60 is also divisible by the sum of its digits, and the number 60 is also the sum of the pair of twin primes, namely $29 + 31 = 60$, and the sum of four consecutive prime numbers $11 + 13 + 17 + 19 = 60$.

## THE NUMBER 61

In the city of Philadelphia, Pennsylvania, the building with the greatest number of floors is One Liberty Place, which has 61 floors.

The number 61 is the 18th prime number and coupled with the number 59 forms a pair of twin primes. The number 61 is also the smallest prime number whose reversal is a square number, $16 = 4^2$.

The number 61 is the sum of the first three primes whose units digit is a 7, as shown here: $7 + 17 + 37 = 61$.

There is a peculiarity about the decimal expansion of the reciprocal of the number 61. It's repeating part consists of 60 digits and where each digit 0 through 9 appears at least six times. Here is the expansion, which then repeats:

$$\frac{1}{61} = 0.\overline{016393442622950819672131147540983606557377049180327868852459}....$$

The number 61 is also the sum of two consecutive squares: $5^2 + 6^2 = 25 + 36 = 61$. However, the number 61 is also the smallest prime number that can be expressed as a square in 1, 2, 3 or 4 ways: $61^2 = 60^2 + 11^2 = 52^2 + 24^2 + 21^2 = 56^2 + 22^2 + 10^2 + 1^2 = 3,721$.

The number 61 can be described as the smallest prime that can be written as the sum of a prime number of primes each taken to a prime power in a prime number (3) of ways:

$$61 = 2^2 + 2^3 + 7^2, \ 61 = 2^2 + 5^2 + 2^5, \text{ and } 61 = 3^2 + 5^2 + 3^3.$$

The number 61 is a factor of $8! + 1$ because $8! + 1 = 40,320 + 1 = 61 \times 661$.

Prime numbers can be generated using the number 61 in a variety of ways:

- Consider $\dfrac{2^{61} - 1}{2 - 1} = 2,305,843,009,213,693,951$ and

  $\dfrac{2^{61} + 1}{2 + 1} = 768,614,336,404,564,651$ are both prime numbers.

- A very large prime number will also result from the sum: $61! + 60! + 59! + 58! + 57! + \ldots + 4! + 3! + 2! + 1!$

- This curious combination of 61 will also generate a prime number: $(61 + 1)^{61} - 61 = 21,671,703,954,068,549,138,435,915,795,531,764,$ $081,214,370,221,990,508,714,369,459,720,504,857,233,908,677,663,$ $822,814,179,625,798,540,682,330,051,$which is a prime number.

- We can also obtain a prime number with the number 61 and its reversal as: $(61 \times 16) + 1 = 977$, which is a prime number.

- Or in another way we can generate a prime number using the following: $\dfrac{61!}{16!} + 1 = 24,259,681,265,945,368,690,635,336,182,592,420,976,187,$ $003,660,648,527,901,491,200,000,000,001$ is a prime number, and when we now shift to multiplication, $61! \times 16! - 1 = 10{,}619{,}994{,}166$, $259{,}276{,}293{,}853{,}401{,}435{,}545{,}009{,}574{,}169{,}949{,}849{,}753{,}082{,}558{,}628$, $187{,}249{,}379{,}905{,}479{,}953{,}612{,}799{,}999{,}999{,}999{,}999{,}999$ is a prime number.

- When we take the reversal of $61^4 = 13{,}845{,}841$, we get, $14{,}854{,}831$, which is a prime number.

The number 61 is embedded in the number 1,098 as follows: $\dfrac{1098}{1+0+9+8} = 61$.

## THE NUMBER 62

The number 62 represents the sum of the faces, vertices, and ages of both an icosahedron $20 + 12 + 30 = 62$, and a dodecahedron $12 + 20 + 30 = 62$. The number 62 is the smallest number that can be expressed as a sum of three distinct squares in two different ways: $62 = 1^2 + 5^2 + 6^2 = 2^2 + 3^2 + 7^2$. The number +62 is the telephone code for the country of Indonesia.

## THE NUMBER 63

The number 63 can be represented as the sum of consecutive powers of 2, as $2^0 + 2^1 + 2^2 + 2^3 + 2^4 + 2^5 = 63$.

The numbers 63 also lends itself to the Kaprekar process (reverse and subtract), which we show as follows: $63 - 36 = 27$, then $72 - 27 = 45$, then $54 - 45 = 09$, then $90 - 09 = 81$, and then $81 - 18 = 63$, which is the number with which we started.

## THE NUMBER 64

Perhaps one of the most notable appearances of the number 64 is the number of squares on a chess or checkerboard.

The number 64 is both a perfect square and a perfect cube as $64 = 4^3 = 8^2$. And it is also a 6th power, as $2^6 = 64$.

The number 64 is also the smallest number that has six powers of 2 factors, namely, 2, 4, 8, 16, 32, and 64.

## THE NUMBER 65

One of the more notable aspects about the number 65 is that it can be expressed as a sum of two squares in two different ways, namely, $65 = 1^2 + 8^2 = 4^2 + 7^2$.

The number 65 can be expressed as the difference of two fourth powers of primes as: $3^4 - 2^4 = 65$.

The number 65 is the only number that can generate a square by adding its reversal, $65 + 56 = 121 = 11^2$, and generate a square by subtracting its reversal, $65 - 56 = 9 = 3^2$.

There is also a Pythagorean triple, where the hypotenuse is 65 and one of its legs is its reversal 56. That triple is (33, 56, 65) because $33^2 + 56^2 = 65^2$. An analogous situation can be seen with the Pythagorean triple (3333, 5656, 6565). Notice the symmetry and relationship that this triple has to the previous one. This kind of course is justified in that $3333^2 + 5656^2 = 43,099,225 = 6565^2$. There are also Pythagorean triples where the two legs have lengths, which are reversals of one another such as the Pythagorean triple (88,209, 90,288, 126,225).

Consider the number $65^{65}$ = **690825216476092085140553869446828608223037872425945418628911729772**99871291049018773300360862776869907975196838378 90625, and take the first 65 places as indicated in bold, which are as follows: 69,082,521,647,609,208,514,055,386,944,682,860,822, 303,787,242,594,541,862,891,172,977, we get a prime number.

We can also get a prime number $6^{65} - 5 = 380,041,719,977,839,666,236, 973,721,680,871,319,659,378,770,968,571$, which is a prime number.

Another way to generate a prime number is $(65!)^2 + 1 = 68,023,740, 289,078,328,950,450,781,972,622,203,792,902,576,953,271,358,034,279, 380,104,027,100,652,464,382,649,659,623,724,446,578,151,412,858,996, 571,534,385,340,563,792,951,822,384,455,180,747,800,000,000,000,000, 000,000,000,000,000,001$, which, despite its inordinate length, is a prime number.

## THE NUMBER 66

The number 66 is both a triangular number and a hexagonal number, as well as being a repdigit number.

The number 66 can be expressed as a sum of triangular numbers in a variety of ways. Here are some ways that it can be suppressed: $21 + 45 = 66, 6 + 15 + 45 = 66, 3 + 6 + 21 + 36 = 66, 1 + 6 + 10 + 21 + 28 = 66, 1 + 3 + 6 + 10 + 10 + 15 + 21 = 66$. Notice, each time the number of addends increased by one.

If one takes the sum of all the divisors of the number 66, a perfect square will be reached, as we can see: $1 + 2 + 3 + 6 + 11 + 22 + 33 + 66 = 144 = 12^2$.

The number 66 is the smallest number where its prime factors $2 \times 3 \times 11 = 66$, has a sum $2 + 3 + 11 = 16$, which is a fourth power as $2^4$.

The number 66 has a peculiar way of being represented as the sum of two prime numbers with a units digit of 3 in two different ways, as shown here: $66 = 13 + 53 = 23 + 43$.

The number 63 is also the smallest number that can be expressed as the sum of two prime numbers in six different ways: $66 = 5 + 61 = 7 + 59 = 13 + 53 = 19 + 47 = 23 + 43 = 29 + 37$.

## THE NUMBER 67

The number 67 is the 19th prime number; however, it cannot be expressed as a sum of two distinct square numbers. It can be expressed as a sum of five consecutive prime numbers, namely, $67 = 7 + 11 + 13 + 17 + 19$.

There is a peculiar way to use the number 9 to represent the number 67 as $\sqrt{\dfrac{9!}{9 \times 9} + 9}$.

Another curious way to represent the number 67 is $2^6 + 2^1 + 2^0 = 26 + 21 + 20 = 67$. Notice how the first representation (with the exponents) relates to the second representation.

## THE NUMBER 68

It is believed that the number 68 is the largest number that can be expressed as the sum of two primes in exactly 2 different ways, namely, $68 = 7 + 61 = 31 + 37$. Although this has not been proved, the conjecture still holds true to this day.

It is possible to construct a 68-sided regular polygon with the traditional construction tools: compasses and unmarked straightedge.

The number 68 is also a happy number. (See page 216)

## THE NUMBER 69

The number 69 is the only number whose square and cube uses all the numbers from 0 to 9 exactly once, as we can see: $69^2 = 4,761$, and $69^3 = 328,509$.

Although the number 69 is not a prime number, we can generate a prime by taking $10^{69} + 69 = 1,000,000,000,000,000,000,000,000,000,000,000,$ $000,000,000,000,000,000,000,000,000,000,069$.

The number 69 can generate all the digits in two calculations: $69^2 = 4,761$, and $69^3 = 328,509$.

## THE NUMBER 70

The number 70 is the 7th pentagonal number.

A prime number can be achieved by taking the reversal of $2^{70} = 1,180,591,620,717,411,303,424$, which is $4,243,031,147,170,261,950,$ $811$, and is a prime number. Moreover, the digits sum of this number happens to also be 70.

Furthermore, the sum of all squares from 1 to 24 is equal to $70^2 = 4,900$.

## THE NUMBER 71

The number 71 is the 20th prime number and the first of a twin prime pair with the number 73. Its reversals 17 and 37 are also prime numbers.

The number 71 also enables peculiarities, such as $7! + 1 = 5,041 = 71^2$. There only two other numbers that have the characteristic that a factorial +1 is equal to the square, and those are the numbers 25 and 121. It should also be noted that 71 is a factor of $9! + 1 = 362,881 = 71 \times 5,111$.

The number 71 is a factor of the sum of all the prime numbers less than 71.

Oftentimes, we stumble on curious coincidences as with the following: $71 - 1 = 1 \times 2 \times 5 \times 7$, and $71 + 1 = 3 \times 4 \times 6$. What is noteworthy here is that we have used all the numerals from 1 through 7 interchangeably.

A true curiosity is when we consider $71^3 = 357,911$; inspecting the digits we find that all the primes between 3 and 11 are provided in order.

We can also arrive at the number 71 by taking the sum of three consecutive prime numbers, namely, $71 = 19 + 23 + 29$.

The number 71 can be expressed in two ways as the sum of consecutive composite numbers, as follows: $71 = 22 + 24 + 25$, and $71 = 35 + 36$.

There are three prime numbers in the 70s, and they are $71, 73, 79$, whose sum is 223, which is a prime number. The sum of the reversals of these three prime numbers, namely, $17, 37, 97$, is 151, which is also a prime number.

## THE NUMBER 72

In a typical 18-hole golf course, 72 is the usual par for the course.

The number 72 can be expressed as the sun of four consecutive prime numbers: $72 = 13 + 17 + 19 + 23$, and also the sum of six consecutive prime numbers: $72 = 5 + 7 + 11 + 13 + 17 + 19$.

The number 72 is the smallest number that can be expressed as a sum of two prime numbers, each of which has a units digit of 1: $72 = 11 + 61$, and $72 = 31 + 41$.

The number 72 is also the smallest number that can be expressed as a difference of two squares of consecutive prime numbers into different ways: $72 = 19^2 - 17^2$, and $72 = 11^2 - 7^2$.

Because we know that the sum of the exterior angles of a regular pentagon is $360°$, then one exterior angle is $\dfrac{360°}{5} = 72°$.

The number 72 is the smallest number whose 5th power is equal to the sum of 5 other 5th powers:

$$72^5 = 19^5 + 43^5 + 46^5 + 47^5 + 67^5.$$

With the following two sets of numbers, the sum of the numbers in each set is 72 and the sum of any three of the numbers in each of the sets will always be a prime number. The two sets are $\{5, 11, 13, 43\}$ and $\{11, 13, 19, 29\}$.

One of the special features of the number 72 is the famous "Rule of 72." It states that—roughly speaking—*money will double in $\dfrac{72}{r}$ years, when it is invested at an annually compounded interest rate of r%.* So, for example, if we invest money at an 8% compounded annual interest rate, it will double its value in $\dfrac{72}{8} = 9$ years.

To investigate why, or if, this really works, we consider the compound interest formula: $A = P\left(1 + \dfrac{r}{100}\right)^n$, where $A$ is the resulting amount of money and $P$ is the principal invested for $n$ interest periods at r% annually.

We need to investigate what happens when $A = 2P$. When we apply the preceding equation we get the following:

$$2 = \left(1 + \frac{r}{100}\right)^n \tag{1}$$

It then follows that $n = \dfrac{\log 2}{\log\left(1 + \dfrac{r}{100}\right)}$. $\tag{2}$

Let us make a table of (rounded) values (Figure 65) from the preceding equation:

| $r$ | $n$ | $n \times r$ |
|---|---|---|
| 1 | 69.66071689 | 69.66071689 |
| 3 | 23.44977225 | 70.34931675 |
| 5 | 14.20669908 | 71.03349541 |
| 7 | 10.24476835 | 71.71337846 |
| 9 | 8.043231727 | 72.38908554 |
| 11 | 6.641884618 | 73.0607308 |
| 13 | 5.671417169 | 73.72842319 |
| 15 | 4.959484455 | 74.39226682 |

**Figure 65**

If we take the arithmetic mean (the average) of the *nr* values we get 72.04092673, which is quite close to 72, and so our "Rule of 72" seems to be a very close estimate for doubling money at an annual interest rate of *r*% for *n* interest periods.

An ambitious reader might try to determine a "rule" for tripling and quadrupling money in a manner similar to the way we dealt with the doubling of money. The preceding equation (2) for *k*-tupling would be

$$n = \frac{\log k}{\log\left(1 + \dfrac{r}{100}\right)},$$ which for *r* = 8, gives the value for *n* = 29.91884022

(log *k*). Thus *nr* = 239.3507218 log *k*, which for *k* = 3 (the tripling effect) gives us *nr* = 114.1821673. We could then say that for tripling money we would have a "rule of 114."

However far one wishes to explore this topic, the important issue here is that the popular "Rule of 72" provides a useful application for investing funds and gives us a curiosity of the number 72 to consider.


## THE NUMBER 73


In New York, the Empire State Building has 73 elevators. In one of the most contested presidential elections in US history, Thomas Jefferson and Aaron Burr had each received 73 electoral votes and were tied for the presidency until eventually Jefferson received more votes than Burr. In the year 2001 the all-time record number of 73 home runs in one season was accomplished by Barry Bonds.

The number 73 is composed of prime digits, whose reversal is also a prime number, 37, and also 73 is the second member of the twin prime with the prime number 71. While we are on 73 and its reversal 37, we can generate a prime number as: $(73 \times 37) + 7^3 + 3^7 + (7 \times 3) + (3 \times 7) = 2{,}593$, which is the 378th prime number.

The number 73, which is the 21st prime number, is the partner twin prime with the number 71, and similarly, when reversed, it also provides a prime number, 37. Moreover, 73 is the 37th prime number.

Additionally, $73! + 1 = 4{,}470{,}115{,}461{,}512{,}684{,}340{,}891{,}257{,}138{,}125{,}051{,}$ $110{,}076{,}800{,}700{,}282{,}905{,}015{,}819{,}080{,}092{,}370{,}422{,}104{,}067{,}183{,}317{,}016{,}$ $903{,}680{,}000{,}000{,}000{,}000{,}001$ is a prime number.

The number 73 can be expressed in a rather symmetric fashion as follows: $73 = 2^{3^1} + 3^{2^1}$, or as $73 = 8^2 + 8^1 + 8^0$.

Here is a cute use for the number 73. The 73rd triangular number is equal to $73 \times 37 = 2,701$. By the way, the formula for finding the $n$th triangular number is $\dfrac{n(n + 1)}{2}$.

## THE NUMBER 75

The number 75 is equal to the sum of the first five pentagonal numbers: $75 = 1 + 5 + 12 + 22 + 35$.

A prime number can be obtained using the number 75 as follows: $2^{75} + 75 = 37,778,931,862,957,161,709,643$, which is a prime number.

The number 75 is the sum of the first five pentagonal numbers: $75 = 1, 5, 12, 22, 35$.

## THE NUMBER 76

In the United States the Declaration of Independence was signed in July 1776 in Philadelphia, and that is probably why the National Basketball Association team is called the Philadelphia 76ers.

When the number 76 is taken to the 2nd power it has curious results: $76^2 = 5,776$, which is a number ending in 76. The only other two-digit number that has this characteristic is 25 because $25^2 = 625$.

## THE NUMBER 77

The number 77 is the sum of the first eight prime numbers: $77 = 2 + 3 + 5 + 7 + 11 + 13 + 17 + 19$.

The number 77 can also be expressed as a sum of three consecutive squares: $77 = 4^2 + 5^2 + 6^2$.

Although it is a very large number, $77! + 1$ is a prime number that is 145, 183,092,028,285,869,634,070,784,086,308,284,983,740,379,224,208,358, 846,781,574,688,061,991,349,156,420,080,065,207,861,248,000,000,000, 000,000,001.

If one takes all the palindromic numbers from 1 through 77 and places them together to form a number as: 12,345,678,911,223,344,556,677, the result is a prime number.

The number 77 will appear as a sum of the digits of a number formed by lots of 7s in the following way: $7^7 + 77^7 + 777^7 + 2 = 170,980,732,128,390,323,351$, whose digit sum is 77.

## THE NUMBER 78

The number 78 is the 12th triangular number.

The earliest phonograph records turned 78 revolutions per minute.

The number 78 is the smallest number with the sum of the reciprocals of the numbers whose sum 78 will be equal to 1. For example: $78 = 2 + 6 + 8 + 10 + 12 + 40$, and as some of the reciprocals:

$$\frac{1}{2} + \frac{1}{6} + \frac{1}{8} + \frac{1}{10} + \frac{1}{12} + \frac{1}{40} = 1$$

## THE NUMBER 79

The only US president, James A. Garfield, who provided an original proof of the Pythagorean theorem died on September 19, 1881, exactly 79 days after he was shot by an assassin on July 2, 1881. The horrific event of the volcano eruption of Mt. Vesuvius occurred in the year 79 CE, burying the cities of Pompeii and Herculaneum.

The 79th triangular number is 3,160, which is comprised of four digits each of which is a triangular number.

The number 79 is the 22nd prime number and whose reversal 97 is also prime. Both of these numbers can be expressed as the sum of three prime numbers that are also reversible primes: $79 = 11 + 31 + 37$, and $97 = 11 + 13 + 73$.

The number 79 is also a happy number because $7^2 + 9^2 = 49 + 81 = 130$, and $1^2 + 3^2 = 10$ and $1^2 + 0^2 = 1$.

The number 79 is the smallest prime where the sum of its digits: $7 + 9 = 16$ is a 4th power, as $2^4$.

The number 79 can be arrived at by finding the sum of the product of its digits and the sum of its digits as follows: $(7 \times 9) + (7 + 9) = 63 + 16 = 79$.

Another clever way of representing the number 79 using only prime numbers is as follows: $2^7 - 7^2 = 79$.

Two prime numbers that are reversals of one another can be generated with the number 79 as follows: $7! + 9! - 79 = 367,841$, which is a prime number, and its reversal 148,763 is also a prime number.

The number 79 is the average of the two prime numbers prior to 79, namely, 73 and 71, and the two prime numbers after it, namely 83 and 89. Symbolically, $\dfrac{73 + 71 + 83 + 89}{4} = 79$.

The number 79 lends itself to a very nice little trick. Select any two two-digit numbers that differ by 2 and with different 10's digits. Then reverse each of the two numbers and subtract the smaller from the larger. Surprisingly, the result will always be 79. For example, consider the selected two numbers 39 and 41 that differ by 2. Subtracting their reversals: $93 - 14 = 79$.

## THE NUMBER 81

The number 81 is both a square number and a 4th power number: $81 = 9^2$ and $81 = 3^4$. From this we can also conclude that the square root of 81 is equal to the sum of its digits, 9.

The sum of the divisors of 81 is $1 + 3 + 9 + 27 + 81 = 121$, which is a square number, $11^2$.

A prime number can be cleverly generated using the number 81 in the following way: $81 \times 2^{81} - 1 = 195{,}845{,}982{,}777{,}569{,}926{,}302{,}400{,}511$, which is a prime number.

The reciprocal of 81, which is $\dfrac{1}{81} = 0.0123456790\ 1234567901234$ 56790 123456790, where the digits are without the number 8.

We can also lead up to the number 81 with a very curious pattern as follows:

$$
\begin{array}{ll}
1 = 3^0 & 1 \\
2 + 3 + 4 = 3^2 & 9 \\
5 + 6 + 7 + 8 + 9 + 10 + 11 + 12 + 13 = 3^4 & = \mathbf{81} \\
14 + 15 + 16 + 17 + 81 + \ldots + 36 + 37 + 38 + 39 + 40 = 3^6 & 729
\end{array}
$$

**Equation 10**

## THE NUMBER 82

The number 82 is equal to the sum of the $(8 \times 2 = 16^{\text{th}})$ prime number and the $(8 + 2 = 10^{\text{th}})$ prime number, in other words $53 + 29 = 82$.

The number 82 is considered a "happy number," which, as has been mentioned earlier, is a number that when you continuously take the sum of the squares of the digits eventually will reach the number 1. For the number 82, the process goes as follows: $8^2 + 2^2 = 68 \rightarrow 6^2 + 8^2 = 100 \rightarrow 1^2 + 0^2 + 0^2 = 1$.

## THE NUMBER 83

The number 83 is the 23rd prime number and can be expressed as a sum of three consecutive prime numbers, namely, $3 + 29 + 31 = 83$, and also can be expressed as the sum of five consecutive prime numbers, namely $11 + 13 + 17 + 19 + 23 = 83$.

The smallest prime number, where the sum of its digits is 83, is the prime number 3,999,998,999.

The square of the number 83 results in a number whose value doesn't change when it's flipped over. That is, $83^2 = 6889$.

The number 83 is equal to the sum of the squares of the first three consecutive odd prime numbers: $83 = 3^2 + 5^2 + 7^2$.

The number 83 can also be obtained by taking the sum of the first three prime numbers whose units digit is a 1, as we see with the following: $83 = 11 + 31 + 41$.

The number 83 is the only two-digit prime number, which is the average of three primes that are reversible prime numbers: $\dfrac{73 + 79 + 97}{3} = 83$.

The number 83 can be expressed as a sum of three triangular numbers as follows: $83 = 10 + 28 + 45$. It is believed that 83 is the largest number that fits this characteristic. However, the number 83 can also be expressed as the sum of 6 triangular numbers: $83 = 3 + 6 + 10 + 15 + 21 + 28$.

The average of all the prime numbers less than or equal to 83 is the reversal of 83, namely 38.

$$(2 + 3 + 5 + 7 + 11 + 13 + 17 + 19 + 23 + 29 + 31 + 37 + 41 + 43 + 47$$
$$+ \ 53 + 59 + 61 + 67 + 71 + 73 + 79 + 83) \div 23 = 874 \div 23 = 38.$$

## THE NUMBER 84

Although the number 84 is not a prime number, it is the sum of two twin prime numbers: $41 + 43 = 84$.

The number 84 is the sum of the first seven triangular numbers: $0 + 1 + 3 + 6 + 10 + 15 + 21 + 28 = 84$.

The number 84 is the smallest number that can be expressed as a sum of two prime numbers in eight different ways, and they are as follows: $(5 + 79)$, $(11 + 73), (13 + 71), (17 + 67), (23 + 61), (31 + 53), (37 + 47), (41 + 43)$.

The number 84 is also the smallest number that can be expressed as the sum of four different primes, where the sum of any three within the set is also a prime number. These four groups of prime numbers are $(5 + 13 + 23 + 43)$, $(11 + 13 + 17 + 43), (11 + 13 + 23 + 37), (13 + 17 + 23 + 31)$.    For    example, if we take the first group listed, we can see that the the sum of any three of the numbers in the group can give us a prime number $5 + 13 + 23 = 41$, $13 + 23 + 43 = 79$, $5 + 23 + 43 = 71$, $5 + 13 + 43 = 61$.

## THE NUMBER 85

The number 85 is the smallest number that can be expressed as a sum of two squares (without using the number 1) in two different ways, namely, $85 = 2^2 + 9^2 = 6^2 + 7^2$.

A pair of twin prime numbers can be generated with the number 85 as follows: $\dfrac{85^{11} - 85}{11} + 1 = 152,130,221,536,012,961,641$ and $\dfrac{85^{11} - 85}{11} - 1 = 152,130,221,536,012,961,639$.

The number 85 is the only composite number that can be written as a sum of three prime numbers that can be reversed and remain prime. That is, $85 = 17 + 31 + 37$, where each of these prime numbers when reversed, 71, 13, 73 are also prime numbers.

## THE NUMBER 86

In the English language the term "you are 86" means that you are being ejected from a restaurant or a bar. There are many definitions as to where this expression came from, but one that seems to be popular is that 86 Bedford Street in Greenwich Village, New York City was the rear door's marking of the Chumley's Bar, and when one had too much alcohol he was told "you're 86" meaning its time for you to leave by the back door. Suffice it to say, being 86ed was not good.

As we noted earlier with the number 82, the number 86 is also a happy number as we can see:
$$8^2 + 6^2 = 64 + 36 = 100 \rightarrow 1^2 + 0^2 + 0^2 = 1.$$
It is also believed that $2^{86} = 77,371,252,455,336,267,181,195,264$ is the largest power of 2 that does not contain a digit 0. When we go to the next power of 2, that is, $2^{87}$ there are three zeros included in the number: $154,742,504,910,672,534,362,390,528$.

## THE NUMBER 87

The number 87 has a few curious summations: it is the sum of the squares of the first four prime numbers: $87 = 2^2 + 3^2 + 5^2 + 7^2$, and the number 87 can also be expressed in terms of factorials as: $87 = 5! - 4! - 3! - 2! - 1!$.

The number 87 can also be expressed as a sum of each of the divisors of the first 10 natural numbers as we show here: $(1)+(1 + 2)+(1 + 3)+(1 + 2 + 4)+(1 + 5)+(1 + 2 + 3 + 6)+(1 + 2 + 4 + 8)+(1 + 3 + 9)+(1 + 2 + 5 + 10)=87.$

The number 87 is a composite number. However, if we place any of these single-digit prime numbers between the two numbers 8 and 7, a prime number is arrived at: 827 is the 144th prime number, 857 is the 148th prime number, and 877 is the 151st prime number.

Another clever pattern to generate prime numbers can emanate from the number 87 in the following way: $87 = 3 \times 29$, now using these two factors, prime numbers can be generated as follows: $87^2 - 3^2 - 29^2 = 6,719$, which is the 867th prime number, and $87^2 + 3^2 + 29^2 = 8,419$, which is the 1,052nd prime number.

## THE NUMBER 88

As was mentioned in the discussion of the number 8, it is considered to be a very lucky number in the Chinese culture; so too, is the number 88 a very lucky number in the Chinese culture. Automobile license plates in China with the numbers 88 are sold for enormously high sums of money.

Not only is the number 88 a repeated-digit number, but also its square does something analogous: $88^2 = 7744$.

Although the number 88 is a composite number, together with powers of the number 91 prime numbers can be generated in a rather unexpected way as follows: $88 + 91 = 179$, $88 + 91^2 = 8,369$, $88 + 91^3 = 753,659$, $88 + 91^4 = 68,575,049$, $88 + 91^5 = 6,240,321,539$, $88 + 91^6 = 567,869,252,129$.

## THE NUMBER 89

The height of the pedestal of the Statue of Liberty is exactly 89 feet.

The number 89 is the 24th prime number and presents the first time that a 2nd prime number, 97, differs by 8.

There is a nice technique for generating palindromic numbers for most starting numbers. One can arrive at a palindromic number by taking the given number and adding it to his reversal in continuing this process until a palindromic number is reached. There are times when this can be accomplished in very few steps. For example, if one were to begin with a number 86, the palindrome would be accomplished in 3 steps: $86 + 68 = 154$, $154 + 451 = 605$, and $605 + 506 = 1,111$, a palindromic number. Were one to do this with the number 89 it would require 24 steps, which is probably the maximum number of steps to achieve the palindrome. Be cautioned! This reversal additions does not produce palindromes for all numbers. Some numbers for which this will not produce a palindromic number are: 196, 691, 788, 887, 1675, 5716, 6347, and 7436.

If we were to test the number 89 to see if it is a happy number, we would use the process described for the number 82 (page 119) and rather than reach the number 1, the number 89 would reappear and create a loop.

Using the digits from 1 to 9 consecutively, and concluding with the number 89, a prime number can be generated as follows: $12 \times 3,456^7 + 89 = 70,663,864,180,791,861,451,948,121$, which is a prime number.

The prime number 89 can also be generated as follows: $8^1 + 9^2 = 89$.

The number 89 joins the number 1 as the only two outcomes that can occur if you take any number and take the sum of the squares of the digits and continue this process. For example, suppose the number 16 is selected as the starting number: $1^2 + 6^2 = 37$, then $3^2 + 7^2 = 58$, then $5^2 + 8^2 = 89$, then $8^2 + 9^2 = 145$, $1^2 + 4^2 + 5^2 = 42$, then $4^2 + 2^2 = 20$, then $2^2 + 0^2 = 4$, then $4^2 = 16$, and if you continue the process you will get back to the number 89 forming a loop.

To get a prime number by squaring or cubing the number 89 we need to reverse the results: $89^2 = 7,921$, and its reversal is 1,297, which is a prime number. Similarly, $89^3 = 704,969$, and its reversal is 969,407, which is also a prime number. This will not work with $89^4$.

Cute patterns for 89 are that $89 = 8 \times 9 + (8 + 9)$, and that $89 = 9 + 8 + 7 + 6 + 5 + 4 + 3 + 2 + 1 + 2 + 3 + 4 + 5 + 6 + 7 + 8 + 9$. Notice the palindromic pattern.

The number 89 is the only prime number where the pattern $2^{89} + 89 = 618,970,019,642,690,137,449,562,201$ is a prime number. Furthermore, $2^{89} - 1 = 618,970,019,642,690,137,449,562,111$ is also a prime number.

There is no single digit that can be placed after the digit 9 in the number 89 that would create a prime number.

Although it may be difficult to count them, there are exactly 1,000 prime numbers between 1 and $89^2 = 7921$.

## THE NUMBER 90

The first thing that most people think about in mathematics when the number 90 comes up is that it is the number of degrees in a right angle.

There is an unusual relationship between two groups of prime numbers in specific order, where the first member of each group, the second member of each group, the third member of each group and so on, will have a sum of 90. The two groups are as follows: (17, 19, 23, 29, 31, 37) and (73, 71, 67, 61, 59, 53). For example, taking the sum of the last member of each of the two groups yields $37 + 53 = 90$, or for another example, the fourth member of each group yields $29 + 61 = 90$.

The number 90 also enables the generation of a prime number in a rather unusual fashion as follows: $\dfrac{90^3 - 1}{90 - 1} = 8191$, which is the 1,028th prime number.

## THE NUMBER 91

The number 91 is a triangular number as well as a hexagonal number.

The number 91 can be reached in a variety of clever ways, for example:

$$91 = 1+2+3+4+5+6+7+8+9+10+11+12+13$$

$$91 = 1+6+12+18+24+30$$

$$91 = 1^2 + 2^2 + 3^2 + 4^2 + 5^2 + 6^2$$

$$91 = 1^2 + 4^2 + 5^2 + 7^2$$

$$91 = 1^2 + 3^2 + 9^2$$

**Equation 11**

The number 91 can also be expressed as a sum of two cubes and the difference of two cubes, as we see with: $91 = 4^3 + 3^3$, and $91 = 6^3 - 5^3$.

The number 91 can be added to various powers of 10 to become prime numbers as follows:

$$10 + 91 = 101,$$
$$100 + 91 = 191,$$
$$1,000 + 91 = 1,091,$$
$$10,000 + 91 = 10,091,$$

which are each prime numbers.

## THE NUMBER 92

The number 92 is a pentagonal number.

## THE NUMBER 94

The number 94 is the smallest even number, aside from 2 and 4, that cannot be expressed as a sum of two numbers in the sequence of twin prime numbers, such as: 3, 5, 7, 11, 13, 17, 19, 29, 31, 41, 43, 57, 59,....

A prime number can also be generated by the number 94 and its reversal 49 as follows: $\dfrac{94!}{49!} + 1 = 178760246022308878857609739788649725932796180031173368595656989947247411200000000001$, which is a prime number.

The factors of the number 94 are 2 and 47, and their sum, 49, is the reverse of 94.

The number 94 is the largest two-digit number, where 1 less than its factorial is a prime number, which can be shown symbolically as: $94! - 1 = 108,736,615,665,674,308,027,365,285,256,786,601,004,186,803,$ $580,182,872,307,497,374,434,045,199,869,417,927,630,229,109,214,583,$ $415,458,560,865,651,202,385,340,530,687,999,999,999,999,999,999,999.$

## THE NUMBER 95

The number 95 represents the number of theses that Martin Luther posted in Wittenberg, Germany on October 31, 1517.

The number 95 can be used to generate a prime number in the following way:

$95^0 + 95^1 + 95^2 + 95^3 + 95^4 + 95^5 + 95^6 = 742,912,017,121$, which is a prime number.

## THE NUMBER 97

The prime number 97 is the largest prime that can be expressed as the sum of equal powers of two consecutive prime numbers: $97 = 2^4 + 3^4$. (It should be noted that the only two consecutive prime numbers are 2 and 3.)

Not only is 97 a prime number, in Shakespeare's sonnet 97 "How like a winter hath my absence been" the seventh word on the seventh line is the word "prime." (Remember: 7 is a prime number.)

> How like a winter hath my absence been
> From thee, the pleasure of the fleeting year!
> What freezings have I felt, what dark days seen!
> What old December's bareness everywhere!
> And yet this time remov'd was summer's time,
> The teeming autumn, big with rich increase,
> Bearing the wanton burthen of the **prime**,

We can insert several 0s between the 2 digits of 97 to generate some prime numbers as follows:

9,907, 9,007, 90,007, 900,007, which are for prime numbers.

The number 97 can be expressed as the sum of an odd number of prime numbers $(29 + 31 + 37 = 97)$ and an even number of composite numbers $(22 + 24 + 25 + 26 = 97)$.

The number 97 can be expressed using the 4 four times as shown: $\frac{4}{4} + 4! \times 4 = 97$.

The number 97 can be expressed as the sum of the first five consecutive odd composite numbers:

$$97 = 9 + 15 + 21 + 25 + 27.$$

The number 97and its reversal 79 are both prime numbers. The next number also with the units digit of 7, is the number 107, which is also a prime number as is its reversal 701. Interestingly, there are 97 prime numbers between 107 and 701.

It is always a challenge to get close to the value of $\pi$ with rational numbers. With the number 97, we can get the value of $\pi$ to 8-place accuracy: $97^{\frac{272}{1087}} \approx 3.14159265\,426788632604403925$.

## THE NUMBER 99

The number 99 is a Kaprekar number, since $99^2 = 9{,}801$ and $98 + 01 = 99$.

The reciprocal of 99 also has a very curious pattern: $\frac{1}{99} = 01010101010101010101010101010101010101010101010101010101010101010.$

## THE NUMBER 100

The temperature of 100° indicates the boiling point of water on the Celsius temperature scale.

The number 100 is the basis for percents (which is where the word *percent* came from: "per hundred").

The number of senators in the US Congress is 100.

The largest denomination of currency in print in the United States is the $100 bill. There are 100 years in a century. The number 100 has been significant in American sports: a football field is 100 yards long, a sprint run is 100 yards or 100 meters, and 100 was the most points scored by one player in a single National Basketball Association game, which was accomplished by Wilt Chamberlain in 1962.

The number 100 can be expressed in a number interesting ways:

- As the square of the sum of the first four natural numbers: $100 = (1 + 2 + 3 + 4)^2$
- As the sum of the cubes of the first four natural numbers: $100 = 1^3 + 2^3 + 3^3 + 4^3$.

- As the sum of the first nine prime numbers: $100 = 2 + 3 + 5 + 7 + 11 + 13 + 17 + 19 + 23$.
- As the sum of 2 to the 6th power and 6 to the 2nd power: $2^6 + 6^2$.
- As using the digits from 1 to 9 to get $100 = 1 + 2 + 3 + 4 + 5 + 6 + 7 + (8 \times 9)$, or using only three operational signs: $100 = 123 - 45 - 67 + 89$.

A motivated reader may develop other ways of reaching the number 100 through arithmetic operations.

## THE NUMBER 101

The number 101 is the 26th prime number and it is also a palindromic number.

The reciprocal of 101, which is $\dfrac{1}{101} \approx 0.00990099009900990099009900990099009900990099$, which is quite unique among prime numbers.

The number 101 can be expressed as a sum of five consecutive prime numbers: $101 = 13 + 17 + 19 + 23 + 29$.

The number 101 can be expressed as $101 = 5! - 4! + 3! - 2! + 1!$

The number 101 can be obtained by taking the sum of the products of all combinations of the first four prime numbers: $101 = (2 \times 3) + (5 \times 2) \times (2 \times 7) + (3 \times 5) + (3 \times 7) + (5 + 7)$.

The number 101 is the largest known prime, which is 1 greater than a power of 10.

There is a clever test to determine if a number is divisible by 101. Although it is a bit cumbersome, it does provide some insight into the number. To do this going from right to left separate the number and pairs of digits then begin by subtracting and adding the digits alternatively. At the end, either a 0 or 101 would indicate that the original number is divisible by 101. For example, let's determine if the number 854,089,043,463 is divisible by 101.

Begin by separating the number into pairs of digits 85 40 89 04 34 63 and then subtract and add pairs alternately: $63 - 34 + 04 - 89 + 40 - 85 = -101$. Therefore, the original number was divisible by 101.

If we list the prime numbers in descending order, beginning with the number 97, we thereby creating this multidigit number: 97 89 83 79 73 71 67 61 59 53 47 43 41 37 31 29 23 19 17 13 11 7 5 3 2. When we subtract 101 from this number, the result is a prime number: $9,789,837,973,716,761,595,347,434,$ $137,312,923,191,713,117,532 - 101 = 9,789,837,973,716,761,595,347,434,$ $137,312,923,191,713,117,431$, which is a prime number.

The number 101 is the first of 21 prime numbers, where, when the center digit is removed, the remaining number is still a prime number. For example, here are a few of these prime numbers which convert to a prime number when the center digit is removed: (101, 11), (103, 13), (107, 17), ... (193, 13), (197, 17), and (199, 19).

## THE NUMBER 102

The number 102 has a very peculiar property, which is as follows: taking the number 102 in parts, we find that the first digit 1 is divisible by 1, and the first 2 digits 10 is divisible by 2, and the first 3 digits (the entire number) 102 is divisible by 3.

The seventh power of the number of the number 102 can be represented as a sum of seventh powers as shown: $102^7 = 114,868,566,764,928 = 12^7 + 35^7 + 53^7 + 58^7 + 64^7 + 83^7 + 85^7 + 90^7$.

## THE NUMBER 103

The number 103 is the 27th prime number and together with the number 101 forms a pair of twin primes.

The number 103 is also happy number because $1^2 + 0^2 + 3^2 = 10 \rightarrow 1^2 + 0^2 = 1$.

A prime number can be generated from the number 103 in a very curious fashion. We will take the reverse of the product of the number 103 and its reverse 301. That is, $103 \times 301 = 31,003$, but its reverse 30,013, is a prime number.

## THE NUMBER 105

The number 105 is a triangular number.

The number 105 fits nicely into the middle of four symmetric prime numbers as follows:

101, 103, **105**, 107, 109.

The number 1 can be represented as a sum of reciprocals of odd numbers with the largest one being 105 as shown here: $1 = \frac{1}{3} + \frac{1}{5} + \frac{1}{7} + \frac{1}{9} + \frac{1}{11} + \frac{1}{33} + \frac{1}{35} + \frac{1}{45} + \frac{1}{55} + \frac{1}{77} + \frac{1}{105}$.

The famous Indian mathematician Srinivas Ramanujan (1887–1920), who discovered many unusual mathematical relationships, also used the number 105 along with the value of $\pi$ to involve prime numbers in a fascinating relationship as follows: $\dfrac{105}{\pi^4} \approx \left(1+\dfrac{1}{2^4}\right)\left(1+\dfrac{1}{3^4}\right)\left(1+\dfrac{1}{5^4}\right)\left(1+\dfrac{1}{7^4}\right)\left(1+\dfrac{1}{11^4}\right)$

$\left(1+\dfrac{1}{13^4}\right)\left(1+\dfrac{1}{17^4}\right)\cdots$

## THE NUMBER 106

We know that the number $106 = 53 \times 2$; yet we can generate a prime number with these three numbers as follows: $106^2 - (53^2 - 2^2) = 8,431$, which is a prime number.

A prime number, albeit a rather long one, can also be generated using the number 106 and its reversal, 601, as follows: $106! - 601 = 114,628,056,373,470,835,453,434,738,414,834,942,870,388,$ $487,424,139,673,389,282,723,476,762,012,382,449,946,252,660,360,871,$ $841,673,476,016,298,287,096,435,143,747,350,528,228,224,302,506,311,$ $679,999,999,999,999,999,999,999,399.$

## THE NUMBER 107

The number 107 is 128th prime number and with the number 109 forms a pair of twin primes.

The reverse number of 107 is number 701, which is also a prime number.

The 107th power of 2 minus 1 yields a prime number:

$$2^{107} - 1 = 162,259,276,829,213,363,391,578,010,288,127.$$

Using the number 107 as an exponent, to get $2^{107} = 162259276829213363391578010288128$, the reverse of which is the prime number $821,882,010,875,193,363,312,928,672,952,261$, as well as $2^{107} - 1 = 162259276829213363391578010288127$, which is also a prime number.

Using the first several prime numbers, the number 107 can be represented as $2 + (3 \times 5 \times 7)$ or $(2 \times 3 \times 5) + (7 \times 11)$.

Because the sum of the digits of the number 107 is 8, we can take the first eight terms of the following and notice that each one results in a prime number: $107 + 2!$, $107 + 3!$, $107 + 4!$, $107 + 5!$, $107 + 6!$, $107 + 7!$, $107 + 8!$, $107 + 9!$. For example, $107 + 9! = 362,987$, which is a prime number.

The sum of the digits of each of the following is 107:

$107^{11} = 21,048,519,522,998,348,950,643$
$107^{13} = 240,984,500,018,808,097,135,911,707$, and
$107^{15} = 240,984,500,018,808,097,135,911,707$, again, each of these three huge numbers have a digit sum equal to 107.

The sum of the digits of the square of the number 107 is equal to the sum of the digits of the square of the number 109, as we can see here: $107^2 = 11,449$, where the digit sum is $1 + 1 + 4 + 4 + 9 = 19$, and for $109^2 = 11,881$, whose digit sum is $1 + 1 + 8 + 8 + 1 = 19$.

## THE NUMBER 108

A major league baseball has 108 stitches. The card game Canasta is played with two decks of cards and four jokers, totaling 108 cards.

The number 108 can be represented as $108 = 1^1 \times 2^2 \times 3^3$.

The number 108 also can relate to the Golden Ratio (see page 6) because $2\sin\left(\dfrac{108°}{2}\right) = 1.61803398875$, which is the Golden Ratio.

## THE NUMBER 109

When carbon dioxide is chilled below $-109°$ Fahrenheit it becomes dry ice. The diameter of the Sun (865,370 miles) is about 109 times the diameter of the Earth (7,926 miles).

The 29th prime number is 109 and is partner with 107 as twin prime numbers.

The number 109 can be expressed as $(1 \times 2) + (3 \times 4) + (5 \times 6) + (7 \times 8) + 9 = 109$, or as $109 = 1 + (2^2 \times 3^3)$.

If one flips the number 109 over, one gets the number 601. Both of these numbers are prime numbers.

The number 109 lends itself to an interesting symmetry: $109 = (10 \times 9) + (10 + 9)$.

If we break up the number 109 as the numbers 10 and 9, we could see that as two sides of a triangle with the third side is 17. This is one of those rare triangles, whose sides are 9, 10, and 17, and where the area, 36, is numerically equal to the perimeter. To get the area of a triangle where only the side lengths are known, we use Heron's formula: $Area = \sqrt{s(s-a)(s-b)(s-c)}$, where $a$, $b$, and $c$ are the triangle's side lengths and $s$ is the semiperimeter. For the given triangle we have:

$$Area = \sqrt{18(18 - 9)(18 - 10)(18 - 17)} = \sqrt{18 \times 9 \times 8 \times 1} = \sqrt{1296} = 36.$$

## THE NUMBER 110

Each of the Twin Towers of the former World Trade Center in New York had 110 stories, which is also the height of the Sears Tower in Chicago. The typical range of electric power in North America is said to be 110 volts; however, the range is between 110 and 120 volts.

The number 110 can be expressed as the sum of three consecutive squares: $110 = 5^2 + 6^2 + 7^2$.

## THE NUMBER 111

As a repunit number, the number 111 is also the smallest palindromic number the sum of whose digits is a prime number.

The number 111 can be represented as a difference of two squares: $20^2 - 17^2 = 111$, as well as the sum of all the composite numbers up to and including the number 17.

The number 111 can be used to generate a prime number in a rather strange way:

$111 \times 2^{111} - 2^{111} - 111 = 285,576,327,219,415,519,569,177,298,107,105,169,$

which is a prime number.

The number 111 can also be obtained by a pair of prime numbers, which are reversals of one another in the following way: $111 = 37 + 73 + 1$.

The number 111 is the sum of each row, column, and diagonal of an unusual magic square, which consists of only prime numbers and the number 1 as shown in Figure 66.

| 31 | 73 | 7 |
|----|----|---|
| 13 | 37 | 61 |
| 67 | 1 | 43 |

**Figure 66**

The number 111 is the sum of each row, column, and diagonal of a 6 × 6 magic square, as shown in Figure 67. Interestingly, the sum of all the numbers in this magic square is 666, which is a number we will cover on page 170.

| 1 | 11 | 31 | 29 | 19 | 20 |
|---|----|----|----|----|----|
| 2 | 22 | 24 | 25 | 8 | 30 |
| 3 | 33 | 26 | 23 | 17 | 9 |
| 34 | 27 | 10 | 12 | 12 | 7 |
| 35 | 14 | 15 | 16 | 18 | 13 |
| 36 | 4 | 5 | 6 | 28 | 32 |

**Figure 67**

## THE NUMBER 113

The number 113 is the smallest three-digit prime number, which can have its digits arranged in any order and still result in a prime number. That is, 131 and 311 are prime numbers.

With the number 113 we can achieve six place accuracy to the value of $\pi$ because

$$\frac{355}{113} \approx \mathbf{3.141592}\ 920353982300884955752212389380530973455\ldots.$$

A rare curiosity occurs when we take $113^2 = 12,769$ and then take the reversal $311^2 = 96,721$, which amazingly is a reversal of the previous square.

Another strange occurrence can be seen when we multiply $113 \times 109,253,795,664,701$. We get the following number, which is a combination of the first 13 numbers listed in order: 12,345,678,910,111,213 better seen as 1 2 3 4 5 6 7 8 9 10 11 12 13.

The number 113 is the sum of two consecutive squares: $113 = 7^2 + 8^2$.

It is possible to create the number 113 as seven different sums using only the first 22 prime numbers in seven different ways:

$$113 = 2 + 19 + 31 + 61 = 3 + 37 + 73 = 5 + 41 + 67 = 7 + 47 + 59$$
$$= 11 + 23 + 79 = 13 + 29 + 71 = 17 + 43 + 53.$$

The sum of the digits of the number 113 and the product of digits are both prime numbers, which are $1 + 1 + 3 = 5$ and $1 \times 1 \times 3 = 3$, respectively. These results are twin primes, 3 and 5.

The number 113 is the smallest prime number, which can be expressed as a sum of 3 fourth powers of prime numbers: $2^4 + 2^4 + 3^4 = 113$.

## THE NUMBER 118

The number 118 is a number that can be written as the sum of three numbers in four different ways, where the product of those three numbers, in each case, is the same, namely 37,800. The three ways to write the sums is: **118** = 15 + 40 + 63, **118** = 14 + 50 + 54, **118** = 18 + 30 + 70, and **118** = 21 + 25 + 72. The product of each group of three numbers is as follows: $15 \times 40 \times 63 = 14 \times 50 \times 54 = 18 \times 30 \times 70 = 21 \times 25 \times 72 = 37,800$.

There are two consecutive composite numbers, 118 and 119, whose reversals, 811 and 911, are both prime numbers.

The sum of the cubes of five consecutive numbers beginning with the number 118 yields a perfect square: $118^3 + 119^3 + 120^3 + 121^3 + 122^3 = 8,643,600 = 2,940^2$.

## THE NUMBER 119

The number 119 can be expressed as a sum of five consecutive primes as follows: $119 = 17 + 19 + 23 + 29 + 31$. Additionally, 119 can also be expressed as a sum of seven primes as follows: $119 = 7 + 11 + 13 + 17 + 19 + 23 + 29$.

The number 119 can also be expressed as follows: $5! - 1 = 120 - 1 = 119$.

The number 119 can be expressed as a product of two primes whose units digit is a 7, namely, $119 = 7 \times 17$.

## THE NUMBER 120

The number 120 is the 15th triangular number and can be reached by getting the sum of the first eight triangular numbers 1, 3, 6, 10, 15, 21, 28, 36. In addition, the first five triangular numbers (1, 3, 6, 10, 15) are factors of the number 120.

The number 120 can also be reached by $5! = 120$, or more simply by the product $4 \times 5 \times 6 = 120$.

Furthermore, the number 120 is the sum of a twin prime pair, $59 + 61 = 120$.

The number 120 can also be expressed as the sum of consecutive prime numbers:

$120 = 23 + 29 + 31 + 37$, or as the sum of consecutive powers of 2, as in the following:

$120 = 2^3 + 2^4 + 2^5 + 2^6$, as well as with consecutive powers of 3, as: $120 = 3^1 + 3^2 + 3^3 + 3^4$.

The number 120 has more divisors (16) than any smaller positive number. These divisors are as follows: 1, 2, 3, 4, 5, 6, 8, 10, 12, 15, 20, 24, 30, 40, 60, 120, yet, surprisingly, the sum of these divisors is $3 \times 120 = 360$.

The number 120 is the smallest three-digit number that can be expressed as a difference of squares of consecutive primes in two different ways: $120 = 17^2 - 13^2 = 31^2 - 29^2$.

The number 120 is the least multiple of 6 that is not preceded or succeeded by a prime number.

The number 120 is the sum of 6 prime numbers $11 + 13 + 17 + 19 + 23 + 37$. Yet, what makes this special is that any five of these six numbers has a sum that is a prime number.

## THE NUMBER 121

The number 121 is a square number $(11^2)$ and also the sum of three consecutive prime numbers, namely $37 + 41 + 43 = 121$.

The number 121 is also the only number that can be expressed as the sum of increased powers of a prime number, such as $121 = 3^0 + 3^1 + 3^2 + 3^3 + 3^4$.

The number 121 is one of three squares that can be expressed as 1 greater than a factorial, as shown here: $5! + 1 = 121$.

The number 121 has another unusual characteristic in that the sum of its digits $(1 + 2 + 1 = 4)$ is equal to the sum of the digits of its prime factorization $(11 \times 11)$, namely, $1 + 1 + 1 + 1 = 4$.

The famous French mathematician Pierre de Fermat (1607–1665) conjectured that 121 is one of two numbers that when increased by 4 will become a perfect cube $(125 = 5^3)$. The other such number is 4 because $8 = 2^3$.

The number 121 can be expressed in three different ways as the sum of a prime number and its reversal. $121 = \mathbf{29} + 92$, $121 = \mathbf{83} + 38$, and $121 = \mathbf{47} + 74$.

## THE NUMBER 125

The number 125 is a perfect cube $(5^3)$ and can be expressed as a sum of two squares in two different ways: $125 = 2^2 + 11^2$ and $125 = 5^2 + 10^2$. When the number 125 is written as 53, it leads to a neat pattern: $5^3 - 2^7 = 5 - 2$.

## THE NUMBER 127

When converting millimeters to inches it is not often that both can be compared without fractions. However, 127 mm is exactly equal to 5 inches.

The number 127 has gained fame for its position disproving a mathematician's conjecture.

The French mathematician Alphonse de Polignac (1817–1890) stated that "every odd number greater than 1 can be expressed as the sum of a power of 2 and a prime number." If we inspect the first few cases, we find that this appears to be a true statement. However, as you will see from the list in Equation 12, it holds true for the odd numbers from 3 through 125 and then is *not* true for 127; after which it continues to hold true again for a while.

The next numbers that fail de Polignac's conjecture are 149, 251, 331, 337, 373, and 509, while another counterexample is 877.

| Odd number | Sum of a power of 2 and a prime number |
|---:|---|
| 3 | $= 2^0 + 2$ |
| 5 | $= 2^1 + 3$ |
| 7 | $= 2^2 + 3$ |
| 9 | $= 2^2 + 5$ |
| 11 | $= 2^3 + 3$ |
| 13 | $= 2^3 + 5$ |
| 15 | $= 2^3 + 7$ |
| 17 | $= 2^2 + 13$ |
| 19 | $= 2^4 + 3$ |
| ... | ... |
| 51 | $= 2^5 + 19$ |
| ... | ... |
| 125 | $= 2^6 + 61$ |
| **127** | **= ?** |
| 129 | $= 2^5 + 97$ |
| 131 | $= 2^7 + 3$ |

**Equation 12**

The number 127 is the smallest prime number that can be written as a sum of the first two or more odd prime numbers, as we can see, where $127 = 3 + 5 + 7 + 11 + 13 + 17 + 19 + 23 + 29$.

The number 127 is equal to the sum: $2^0 + 2^1 + 2^2 + 2^3 + 2^4 + 2^5 + 2^6$, or as the sum of the first three odd factorials: $127 = 1! + 3! + 5!$.

The number 127 can also be expressed in terms of a pair of twin prime numbers as $127 = 5! + 7$.

The number 127 is a prime number that is sandwiched between two other primes relatively far apart and are formed by the same digits namely, 113 and 131.

The number 127 is can also be seen as the average of four consecutive prime numbers: $\dfrac{113 + 127 + 131 + 137}{4} = 127$.

We can also arrive at the number 127 in a rather strange way if we take the sum of the reversals of the first five squared numbers $1 + 4 + 9 + 61 + 52 = 127$, and we can also attain 127 by taking the sum of the reversals of the first four cubes $1 + 8 + 72 + 46 = 127$.

The 127th prime number is 709, and the 127th happy number is 907. Not only are they reversals of one another, but they are also both prime numbers.


## THE NUMBER 128


The number 128 is a composite number; however, the sum of the digits, 11, is a prime number.

The number 128 is a power of 2, as $128 = 2^7$, and it is the only 7th power that is three digits long.

Also, each of the three digits of the number 128 is a power of 2.

The number 128 has eight divisors, namely, 1, 2, 4, 8, 16, 32, 64, 128. And 128 is divisible by 8, which is the number of divisors of 128.

The number 128 can also be expressed in terms of the first three prime numbers: $128 = 2! + 3! + 5!$.


## THE NUMBER 129


The number 129 is equal to the sum of the first 10 prime numbers, as we see with: $2 + 3 + 5 + 7 + 11 + 13 + 17 + 19 + 23 + 29 = 129$.

The number 129 is also a happy number, which we can see because $1^2 + 2^2 + 9^2 = 86$, and $8^2 + 6^2 = 100$, and $1^2 + 0^2 + 0^2 = 1$, which verifies the happy number status because the result is 1.

The number 129 is also the smallest number that can be expresses sum of three squares in four different ways: $129 = 1^2 + 8^2 + 8^2$, and $129 = 2^2 + 2^2 \; 11^2$, $129 = 10^2 + 5^2 + 2^2$, and $129 = 4^2 + 7^2 + 8^2$.

When the number 129 is factored, the result is two prime numbers because $3 \times 43 = 129$. Now if we take these two factors and place them together to get the number 343, we get a prime number cubed, namely, $7^3$.

## THE NUMBER 130

The number 130 is the only integer that is equal to the sum of the squares of its first four factors:

$$130 = 1^2 + 2^2 + 5^2 + 10^2.$$

Recall the famous Fibonacci sequence: 1, 1, 2, 3, 5, 8, 13, 21, 34, and so forth. Now we can get 130, if we take the sum of the factorials of the first five terms: $1! + 1! + 2! + 3! + 5! = 130$.

## THE NUMBER 131

The number 131 is a prime number and, as we mentioned earlier (when we discussed the number 113), can have its digits rearranged in any order and still remain a prime number. Furthermore, the sum of the digits, 5, is also a prime number.

The number 131 can also be expressed as a sum of three consecutive prime numbers as follows: $131 = 41 + 43 + 47$. Perhaps even more impressively, the number 131 can be expressed as follows:

$$131 = \left(1^0 + 3^0 + 1^0\right) + \left(1^1 + 3^1 + 1^1\right) + \left(1^2 + 3^2 + 1^2\right)$$
$$+ \left(1^3 + 3^3 + 1^3\right) + \left(1^4 + 3^4 + 1^4\right).$$

The sum of the composite numbers from 2 through 19, including the number 19, is equal to 131, as we can see here: $131 = 4 + 6 + 8 + 9 + 10 + 12 + 14 + 15 + 16 + 18 + 19$. A further curiosity regarding the number 131 is that it is the sum of the odd numbers that have no even digits from 1 to 31, as shown here: $131 = 1 + 3 + 5 + 7 + 9 + 11 + 13 + 15 + 17 + 19 + 31$.

If we place a 7 on both sides of 131 we get the number 71,317, which is a palindrome and also a prime number. The same thing is true we tag a 1 to each side of the number 131, to get 11,311, which again, is both a palindrome and a prime number.

Factorials can also be used to generate the number 131 as follows: $(2! + 1) + (3! + 1) + (5! + 1)$.

Once again, we encounter the Fibonacci numbers $(0, 1, 1, 2, 3, 5, 8, 13, 21,$ 34, etc.) to represent the number 131 with factorials of the first six members: $131 = 0! + 1! + 1! + 2! + 3! + 5!$.

The reciprocal of the number 131 also offers a surprise. $\dfrac{1}{131} \approx 0.0076335$ 87786259541984732824427480916030534351145038167938931297709923664 1221374045801526717557251908396946564885496183206106870229... and these 130 digits then repeat indefinitely.

In mathematics we sometimes look for coincidences. Consider the first few digits of the value of $\pi$, which are 3.14159. if we break these up in pairs and add those numbers: $31 + 41 + 59$, we get a sum of 131.

## THE NUMBER 132

The number 132 is the smallest number that can be expressed as the sum of the two-digit numbers formed by all variations of its three digits. That is, $12 + 13 + 21 + 23 + 31 + 32 = 132$.

The number 132 and its reversal 231 are both divisible by 11. Not only is this highly unusual, but the two quotients upon dividing by 11 are 12 and 21, respectively, which are also reversals of one another. This makes this even more unusual.

## THE NUMBER 135

The number 135 can be expressed in a rather strange fashion: $135 = (1 + 3 + 5) \times (1 \times 3 \times 5)$, or as $135 = 1^1 + 3^2 + 5^3$. Furthermore, the number 135 can be expressed as the sum of consecutive squares: $135 = 3^2 + 4^2 + 5^2 + 6^2 + 7^2$.

Coincidentally, it has also been established that there are 135 prime numbers between 1,000 and 2,000.

## THE NUMBER 136

The number 136 can be expressed as a sum of the cubes of its digits and then of that sum to get back to the original number, as we can see here: $1^3 + 3^3 + 6^3 = 244$ and $2^3 + 4^3 + 4^3 = 136$.

The number 136 is equal to the sum of the first 16 positive integers, namely,

$$1 + 2 + 3 + 4 + 5 + 6 + 7 + 8 + 9 + 10 + 11 + 12 + 13 + 14 + 15 + 16 = 136.$$

## THE NUMBER 137

The 137th St. subway station on the New York City's first subway line, the IRT line (today known as the number #1 train), is the station for the City College of New York of the City University of New York, which was the first free public college in the United States, as it was founded in 1847.

The number 137 is the first of two twin primes with its partner 139. They can also partner in a peculiar way and still produce a prime number as we see in the following: $\dfrac{137! - 1}{139}$ is a prime number.

The number 137 can be expressed sum of two squares: $137 = 4^2 + 11^2$, as well as the sum of the products of the first consecutive odd numbers as: $137 = (1 \times 3) + (5 \times 7) + (9 \times 11)$.

The reciprocal of 137, namely, $\dfrac{1}{137} = 0.0072992700$, produces a repeating palindromic decimal.

While considering palindromes, the number 137 is a factor of the palindrome 123,456,787,654,321 because it is equal to $137 \times 901,144,435,433$. It is also a factor of the repunits number 11,111,111 because $137 \times 81,103 = 11,111,111$.

The number 137 is the only three-digit prime number, which is a factor of a number whose digits follow the pattern: xxxxxyyxxyyxxxxx. For example, $4444455445544444 = 32,441,280,624,412 \times 137$. A motivated reader might want to try this with other such formed numbers.

The number 137 can be expressed as a sums of the products of pairs of consecutive odd numbers, as we can see the following: $(1 \times 3) + (5 \times 7) + (9 \times 11) = 137$.

The number 137 is the smallest three-non-zero-digit prime number, where all two-digit permutations also form prime numbers as the following: 13, 17, 31, 37, 71, 73, which are all prime numbers.

The first few digits of $\pi$ are 3.141592. If we square each of these digits and then add, we will get the number 137, as we see here: $137 = 3^2 + 1^2 + 4^2 + 1^2 + 5^2 + 9^2 + 2^2$.

The sum of the squares of the digits of the number 137 is equal to the prime number $59 = 1^2 + 3^2 + 7^2$. You might notice that in this calculation each of the odd digits was used exactly once.

The number 137 is the only three-digit prime number whose reciprocal produces an infinite periodic palindrome: $\dfrac{1}{137} = 0.00\overline{7299270}07299270072992700....$

## THE NUMBER 138

The number 138 is not a prime number, but it can be expressed as the sum of four consecutive prime numbers, such as $29 + 31 + 37 + 41 = 138$.

## THE NUMBER 139

The number 139 is the first prime number that differs from its next prime number, 149, by ten numbers.

The product of the digits of the prime number 139 is a cube of a prime number, namely, $1 \times 3 \times 9 = 27 = 3^3$.

Similar to the number 138, the prime number 139 can be expressed as a sum of five consecutive prime numbers, such as $139 = 19 + 23 + 29 + 31 + 37$.

Once again, we have a happy number in the number 139 because $1^2 + 3^2 + 9^2 = 91$, and $9^2 + 1^2 = 82$, and $8^2 + 2^2 = 68$, and $6^2 + 8^2 = 100$, and $1^2 + 0^2 + 0^2 = 1$, thus qualifying as a happy number.

The number 139 can be expressed by using the first nine digits in reverse order as follows:

$$(9 \times 8) + (7 \times 6) + (5 \times 4) + (3 \times 2) - 1 = 139.$$

The number 139 can be used as follows as exponents to generate other prime numbers: $2^1 + 2^3 + 2^9 - 1 = 521$, which is a prime number. Also $2^1 + 2^3 + 2^9 + 1 = 523$, which is also a prime number. We can extend this as follows: $2^1 + 2^3 + 2^9 + 139^2 = 19,843$, which is a prime number as well.

The number 139 used as an exponent can lead to a prime number: $3^{139} + 2 = 2,088,595,827,392,656,793,085,408,064,780,643,444,068,898,$ $148,936,888,424,953,199,350,269$, which is a prime number.

## THE NUMBER 140

The number 140 can be expressed as a sum of the squares of the first seven natural numbers: $1^2 + 2^2 + 3^2 + 4^2 + 5^2 + 6^2 + 7^2 = 140$.

## THE NUMBER 143

The number 143 can be expressed as the sum of seven consecutive prime numbers as: $11 + 13 + 17 + 19 + 23 + 29 + 31 = 143$.

The number 143 can be used in a rather strange way to generate a prime number as follows: $\dfrac{143143}{143^2} = \dfrac{143143}{20449} = 7$.

The difference between $3^4 + 4^4 + 5^4 + 6^4$ and $7^4$ is 143 because $2,258 - 2,401 = 143$.

## THE NUMBER 144

As the 12th number in the Fibonacci sequence, the number 144 is the largest in the series, which is also a perfect square, $12^2$, and divisible by the first two primes in the series, namely, 2 and 3.

Curiously, we know that $144 = 12^2$, and if you reverse 144 the square will also be reversed $441 = 21^2$.

The number 144 is also the sum of the twin prime pair, namely, $71 + 73 = 144$.

An interesting relationship between $144^5$ and the sum of some fifth power terms is as follows:

$$144^5 = 61,917,364,224 = 27^5 + 84^5 + 110^5 + 133^5.$$

## THE NUMBER 145

The number 145 can be expressed in a number of ways, such as: $3^4 + 4^3 = 145$, or as $145 = 1! + 4! + 5!$

The number 145 can also be expressed as the sum of two squares in two different ways as follows:

$$145 = 1^2 + 12^2 = 8^2 + 9^2.$$

## THE NUMBER 149

The number 149 is the 35th prime number, which together with 151 forms a twin pair of prime numbers. By the way, the reversal of 149 is also a prime number 941.

If we carefully inspect the number 149, we will notice four perfect squares can be found: 1, 4, 9, and 49.

We can use a clever little trick to create new prime numbers from the number 149 by placing a 0 between pairs of digits to get the numbers 1049, and 1409, which are both prime numbers.

The number 149 can be expressed as the squares of 3 consecutive numbers: $149 = 6^2 + 7^2 + 8^2$.

The number 149 lends itself to an unusual prime number construction, where the number is formed by the number of digits, which is determined by the value of the digit. Here the number has one digit of 1, four digits of 4, and nine digits of 9, resulting in the prime number 14,444,999,999,999.

It is unusual to create a relationship as the one offered here. The number 149 has the product of its digits, $1 \times 4 \times 9 = 36$, which is equal to the sum of the square roots of the cubes of the digits. In other words, $\sqrt{1^3} + \sqrt{4^3} + \sqrt{9^3} = \sqrt{1} + \sqrt{64} + \sqrt{729} = 1 + 8 + 27 = 36$.

## THE NUMBER 150

The number 150 can be expressed as the sum of 8 consecutive prime numbers as shown here: $7 + 11 + 13 + 17 + 19 + 23 + 29 + 31 = 150$.

The number 150 is divisible by the sum of its digits, namely, $150 \div 6 = 25$.

Twin prime numbers occur rather often at the start of our list of prime numbers. The largest gap between pairs of twin primes between the numbers 1 and 1,000 is 150 numbers and occurs between the pairs of twin primes (659, 661) and (809, 811).

## THE NUMBER 151

The height of the Statue of Liberty from the base to the torch is 1 inch more than 151 feet.

The number 151 is a prime number, and it enables generating prime numbers in a rather curious fashion. Using the number 151 we can get the following prime numbers:

- $151 - 1! \times 5! \times 1! = 31$, which is a prime number.
- $151 + 1! \times 5! \times 1! = 327$, which is also a prime number.
- $151 \times 1! \times 5! \times 1! - 1 = 18,119$, is another prime number.
- $151 \times 1! \times 5! \times 1! + 1 = 18,121$, is also a prime number.

## THE NUMBER 152

Public School No. 152 in upper Manhattan has produced some very significant people.

The number 152 is the largest even number that can be expressed uniquely as the sum of two prime numbers in four different ways: $109 + 43$, $113 + 39$, $139 + 13$, and $149 + 3$.

The number 152 is the smallest number that can be expressed as a sum of the cubes of two different odd prime numbers: $152 = 3^3 + 5^3$.

## THE NUMBER 153

The number 153 is the 17th triangular number, which is equal to the sum of the first five factorials: $1! + 2! + 3! + 4! + 5! = 153$, and also the sum of the first 17 integers: $1 + 2 + 3 + 4 + 5 + 6 + 7 + 8 + 9 + 10 + 11 + 12 + 13 + 14 + 15 + 16 + 17 = 153$.

The number 153 is the smallest three-digit number that can be expressed as a sum of the cubes of its digits as $153 = 1^3 + 5^3 + 3^3$. There are three other numbers that share this unique characteristic of being equal to the sum of the cubes of their digits, and they are 370, 371, and 407.

There is a clever trick that can be done with the number 153. Select any integer that is divisible by 3 and then take the sum of the cubes of its digits and continue to do that with each result and you will end up with the number 153. For example, suppose we begin with the number 297. The sum of the cubes of its digits is: $2^3 + 9^3 + 7^3 = 1,080$, continuing this process:

$$1^3 + 0^3 + 8^3 + 0^3 = 513, \text{ then } 5^3 + 1^3 + 3^3 = 153.$$

## THE NUMBER 157

The number 157 is the 37th prime number, preceded by the prime number 151 and succeeded by the prime number 163. The curious aspect is that the average of these three prime numbers is, in fact, 157 as $\dfrac{151 + 157 + 163}{3} = 157$.

Another curiosity embracing the number 157 is that its square and that of the next number 158 use the same digits: $157^2 = 24,649$, and $158^2 = 24,964$.

It is always curious when the value of $\pi$ gets very closely involved with rational numbers. We see this with the number 157, where we find that $5^\pi \approx 156.99254530886589\ldots \approx 157$.

There are not many calculations that use all the digits exactly once without using 0. One such involves the number 157 because $157 = \dfrac{4396}{28}$.

## THE NUMBER 158

The number 158 is the smallest number which when added to its reverse, 851, yields a nonpalindromic prime number, namely, 1009.

As we know, the expression 100! indicates the product of all the numbers from 1 to 100 including both ends. There are 158 digits to this product, which is:

100! = 93,326,215,443,944,152,681,699,238,856,266,700,490,715,968,264,

381,621,468,592,963,895,217,599,993,229,915,608,941,463,976,156,518,

286,253,697,920,827,223,758,251,185,210,916,864,000,000,000,000,000,

000,000,000.

## THE NUMBER 160

The number 160 can be expressed as the sum of the first 11 prime numbers as follows:

$$160 = 2 + 3 + 5 + 7 + 11 + 13 + 17 + 19 + 23 + 29 + 31.$$

The number 160 can also be expressed as the sum of the cubes of the first three prime numbers:

$$160 = 2^3 + 3^3 + 5^3.$$

## THE NUMBER 167

The number 167 is a prime number and can be expressed in terms of the first three prime numbers as $(2 \times 3^4) + 5 = 167$.

The reciprocal of the number 167 results in a repeating decimal with 166 places as shown here: $\dfrac{1}{167} = 0.005988023952095808383233532934131736526 9$

461077844311377245508982035928143712574850299401197604790419161 6 766467065868263473053892215568862275449101796407185628742514 97....

We encountered an analogous situation when we considered the reciprocal of the number 131, which also had a decimal length of 1 less that the number.

The number 167 can also be used to express an equality using its digits as:

$$1 \times 6 \times 7 = 3 \times (1 + 6 + 7)$$

## THE NUMBER 168

The first two perfect numbers in our number system are 6 and 28 since each of these numbers is equal to the sum of their factors. However, the number 168 is equal to the product of these two perfect numbers ($6 \times 28 = 168$).

There are 168 prime numbers less than 1,000, that is, between 2 and 997.

The number 168 can also be expressed as a sum of four consecutive prime numbers, namely, $168 = 37 + 41 + 43 + 47$.

## THE NUMBER 169

The number 169 is a square number ($169 = 13^2$) such that when reversed is also a square number ($961 = 31^2$), yet when flipped over it is 691, which is a prime number.

A curiosity involving the number 169 occurs when we replace the number with one that repeats the digit 6 six times and the digit 9 nine times and whereby the result is a prime number: 1,666,666,999,999,999. Analogously, the same is true with the reverse number 961, where 9,999,999,996,666,661 becomes a prime number, and also when the flipped number 691 is treated analogously it becomes a prime number: 6,666,669,999,999,991.

The number 169 can also be expressed as the sum of squares as follows:

$$169 = 13^2 = 5^2 + 12^2 = 3^2 + 4^2 + 12^2 = 4^2 + 5^2 + 8^2 + 8^2.$$

Another curiosity involving the number 169 can be seen through the use of factorials as follows:

$1! + 6! + 9! = 363,601$, repeating this process: $3! + 6! + 3! + 6! + 0! + 1! = 1,454$, and when we repeat this process once more, we get $1! + 4! + 5! + 4! = 169$, which is the number we started with.

## THE NUMBER 170

The famous Roman mathematician Claudius Ptolemy who was born in the year 100 CE died in the year 170 CE.

The number 170 can be expressed in a rather curious fashion as follows: $170 = (2^2 + 3)^2 + (3^2 + 2)^2$.

## THE NUMBER 173

The number 173 is equidistant between two other prime numbers, namely, 167 and 179.

The number 173 can be expressed as the sum of three consecutive prime numbers, namely, $53 + 59 + 61 = 173$.

The number 173 can also be expressed as the sum of the squares of 2 other prime numbers as follows: $2^2 + 13^2 = 173$.

## THE NUMBER 175

The number 175 can be obtained by consecutively raising each of its digits to one higher power as shown: $175 = 1^1 + 7^2 + 5^3$. Other numbers that share this property are 135, 518, 598, and 1,306.

Although the number 175 is a composite number, when the digits are arranged in ascending order to get 157 and in descending order to get 751, each of these numbers is a prime number.

## THE NUMBER 177

The number 177 is the sum of all the prime numbers from 2 to 47, that is, $2 + 3 + 5 + 7 + 11 + 13 + 17 + 19 + 23 + 29 + 31 + 37 + 41 + 43 + 47 = 177$. The 15th $(1 + 7 + 7)$ prime number is 47, which was the last prime we used in this sum.

A *prime* magic square is one in which all the elements in the square are prime numbers. And the smallest sum of rows, columns, and diagonals of such a $3 \times 3$ magic square is 177, as you can verify in the magic square in Figure 68.

| 17 | 89 | 71 |
|-----|-----|-----|
| 113 | 59 | 5 |
| 47 | 29 | 101 |

**Figure 68**

## THE NUMBER 179

The number 179 is a prime number and can be generated by its digits as follows:

$$179 = (17 \times 9) + (17 + 9).$$

A prime number can be generated with the number 179 and its reversal 971 as follows:

$(179 \times 971) + 179 + 971 = 174,959$, which is a prime number.

The sum of all double-digit prime numbers with consecutive digits in ascending order is 179 as shown here: $179 = 23 + 67 + 89$.

## THE NUMBER 180

On the Fahrenheit temperature scale the difference between the boiling point 212 degrees and the freezing point 32 degrees is equal to 180. The number 180 is well known as it represents the sum of the angle measures (in degrees) of a plain triangle. It is also the measure of a semicircle, 180 degrees.

The number 180 can be expressed as the sum of six consecutive prime numbers, $180 = 19 + 23 + 29 + 31 + 37 + 41$, or it can be expressed as a sum of eight consecutive prime numbers, namely, $180 = 11 + 13 + 17 + 19 + 23 + 29 + 31 + 37$.

The number 180 can also be expressed as the product of the squares of the first two prime numbers times the sum of the same two prime numbers: $180 = (2^2 \times 3^2) \times (2 + 3)$.

It is curious that $180^3 = 6^3 + 7^3 + 8^3 + 9^3 + \ldots + 66^3 + 68^3 + 69^3 = 5,832,000$.

## THE NUMBER 181

The number 181 is the ninth palindromic prime and if you consider the first two digits 18 and the second two digits 81, the number 9 is a factor of each one.

The number 181 is a twin prime with the number 179.

The number 181 can be expressed as the sum of two consecutive square numbers $(181 = 9^2 + 10^2)$ as well as the difference of two consecutive square numbers $(181 = 91^2 - 90^2)$.

The number 181 can also be expressed as the sum of five consecutive prime numbers, namely, $181 = 29 + 31 + 37 + 41 + 43$, as well as the sum of the squares of three consecutive composite numbers: $181 = 6^2 + 8^2 + 9^2$.

Every natural number greater than 181 can be expressed as a sum of cubes of the first two prime numbers, namely 2 and 3. For example,

$$182 = 2^3 + 2^3 + 2^3 + 2^3 + 2^3 + 2^3 + 2^3 + 2^3 + 2^3 + 2^3 + 2^3 + 2^3 + 2^3$$
$$+ 2^3 + 2^3 + 2^3 + 3^3 + 3^3.$$

The number 181 is equal to the sum of the digits of the first 23 prime numbers, namely, from 2 to 83, shown here: $2 + 3 + 5 + 7 + 1 + 1 + 1 + 3 + 1$ $+ 7 + 1 + 9 + 2 + 3 + 2 + 9 + 3 + 1 + 3 + 7 + 4 + 1 + 4 + 3 + 4 + 7 + 5 + 3$ $+ 5 + 9 + 6 + 1 + 6 + 7 + 7 + 1 + 7 + 3 + 7 + 9 + 8 + 3$.

## THE NUMBER 188

The number 188 is the largest even number that can be expressed as the sum of two different primes in exactly five ways. These are: $7 + 181, 31 + 157, 37 + 151, 57 + 131,$ and $79 + 109$.

The number 188 can be expressed as the difference of two squares, as $48^2 - 46^2 = 188$.

The number 188 is a happy number because $1^2 + 8^2 + 8^2 = 129$, and $1^2 + 2^2 + 9^2 = 86$, and then $8^2 + 6^2 = 100$, and finally, $1^2 + 0^2 + 0^2 = 1$, thus, a happy number is achieved.

## THE NUMBER 190

The number 190 is a triangular number, a hexagonal number, and also a happy number as we can see from the following: $1^2 + 9^2 + 0^2 = 82$, $8^2 + 2^2 = 68$, and $6^2 + 8^2 = 100$, which then leads to 1, making 190 a happy number.

When written in Roman numerals, the number 190 is palindromic as it is written CXC.

## THE NUMBER 191

In the collection of the American coins, that is, the penny, the nickel, the dime, the quarter, the half–dollar, and the dollar coin, the total value is 191 cents.

The palindromic prime number 191 has an extra nice feature, which is that the sum of its digits results in another palindromic prime number, 11.

The number 191 is the first of four consecutive prime numbers, 191, 193, 197, and 199, where the sum of the digits of each of these four prime numbers is 11, 13, 17, and 19, respectively, and are consecutive prime numbers.

If we inspect the square of 191, we get $191^2 = 36,481$, where, curiously, the two end digits on either end form a perfect square, 36 and 81, and the middle number, 4, is also perfect square.

The number 191 can assist in creating prime numbers from the factorials of five consecutive prime numbers as follows:

$2! + 191 = 193$, $3! + 191 = 197$, $5! + 191 = 311$, $7! + 191 = 5,231$, and $11! + 191 = 39,916,991$. Each of these five numbers is a prime number.

The number 191 can also be expressed as the difference of the product of two consecutive primes and the sum of those same two consecutive primes as follows: $(13 \times 17) - (13 + 17) = 191$.

## THE NUMBER 192

The number 192 is a curious number in that all of its factors are 2s and a 3, in that $2 \times 2 \times 2 \times 2 \times 2 \times 2 \times 3 = 192$. This can be expressed more attractively, $192 = 4 \times 6 \times 8$.

The number 192 can be expressed as the sum of 10 consecutive prime numbers, as $5 + 7 + 11 + 13 + 17 + 19 + 23 + 29 + 31 + 37 = 192$.

The number 192 also is the smallest number with 14 divisors, namely, $1, 2, 3, 4, 6, 8, 12, 16, 24, 32, 48, 64, 96$, and $192$. Furthermore, the number 192 is also divisible by the sum of its digits, namely, $1 + 9 + 2 = 12$, and then $\dfrac{192}{12} = 16$.

The number 192 is also a happy number because $1^2 + 9^2 + 2^2 = 86$, and $8^2 + 6^2 = 100$, which then leads to 1, making 190 a happy number.

Consider the addition: $192 + 384 = 576$. You may ask, what is so special about this addition? Look at the outside digits (bold): **192** + **384** = **576**. They are in numerical sequence left to right $(1, 2, 3, 4, 5, 6)$ and then reversing to get the rest of the 9 digits $(7, 8, 9)$. You might have also noticed that the 3 numbers we used here are:

$192 = 1 \times 192,$

$384 = 2 \times 192,$

$576 = 3 \times 192.$

## THE NUMBER 193

The number 193 is a prime number as well as a happy number, which we can show by the following: $1^2 + 9^2 + 3^2 = 91$, and $9^2 + 1^2 = 82$, and $8^2 + 2^2 = 68$, and $6^2 + 8^2 = 100$, which then leads to 1, making 193 a happy number.

The number 193 can be expressed as a sum of two squares as well as the difference of two squares as follows: $193 = 12^2 + 7^2$, and $193 = 97^2 - 96^2$.

Curiously, the number 193 can be expressed as the sum of products of the first three twin pair primes as follows: $(3 \times 5) + (5 \times 7) + (11 \times 13) = 193$.

The number 193 can be obtained using the first four prime numbers as follows:

$$(2 \times 3 \times 5 \times 7) - (2 + 3 + 5 + 7) = 193.$$

Although the Swiss mathematician Leonhard Euler's $e$ is an irrational number, which can be represented as $e = \left(1 + \dfrac{1}{n}\right)^n$ as $n$ gets larger and larger where it gets closer and closer to the value of $e \approx 2.71828182845904523536028747135\ldots$. However, the number 193 also allows us to get rather close to $e$ with the following fraction: $\dfrac{193}{71} \approx 2.71830985915492957746478873239436619$.

The number 193 lends itself to an interesting manipulation: if we square 193, we get 37,249. If we tag onto this square the reverse number, 94,273, we get the palindrome 3,724,994,273, which is the product of one palindromic prime number 3,721,273, and one palindromic non-prime number 1,001.

## THE NUMBER 196

The composite number 196 has an interesting partner in the number 169, in that $196 = 14^2$ and $169 = 13^2$.

However, the real fame that the number 196 enjoys in mathematics is that it is the smallest number that through the reversal process does not produce a palindromic number. (At least to date it has not produced a palindromic number.) Let's see what this all means. We must first make a slight diversion into palindromic numbers.

We already know that palindromic numbers are those that read the same in both directions. This leads us to consider that dates can be a source for some symmetric inspection. For example, the year 2002 is a palindrome, as is 1991.[28] There were several dates in October 2001 that appeared as palindromes when written in American style: 10/1/01, or 10/22/01, and others. In February, Europeans had the ultimate palindromic moment at 8:02 P.M. on February 20, 2002, because they would have written it as 20.02, 20-02-2002. It is a bit thought provoking to come up with other palindromic dates.

As we have seen earlier, the first four powers of 11 are palindromic numbers:

$$11^1 = 11$$
$$11^2 = 121$$
$$11^3 = 1331$$
$$11^4 = 14641$$

As we have experienced often till now, a palindromic number can either be a prime number or a composite number. For example, 151 is a prime palindrome and 171 is a composite palindrome. Yet, with the exception of 11, a palindromic prime must have an odd number of digits.

Perhaps most interesting is to see how a palindromic number can be generated from any given number. All we need do is to add continually a number to its reversal (i.e., the number written in the reverse order of digits) until a palindrome results. For example, a palindrome can be reached with a single addition such as with the starting number 23, we have 23 + 32 = 55, a palindrome.

Or it might take two steps to reach a palindrome, such as with the starting number 75:

75 + 57 = 132, 132 + 231 = 363, a palindrome.
Or it might take three steps, such as with the starting number 86:
86 + 68 = 154, 154 + 451 = 605, 605 + 506 = 1,111, a palindrome.

The starting number 97 will require 6 steps to reach a palindrome, while the number 98 will require 24 steps to reach the palindromic number 8,813,200,023,188.

Now we come to the number 196. Many attempts have been made to use this rule of reverse-and-add with the number 196 to eventually reach a palindromic number. So far this has never been accomplished. As a matter of fact, the number 196 is the smallest number that (to the best of our current knowledge) does not generate a palindromic number through the reverse-and-add rule. The next few numbers that do not generate a palindromic number with this procedure are: 295, 394, 493, 592, 689, 691, 788, 790, 879, 887, ... .

There are some lovely patterns when dealing with palindromic numbers. For example, numbers that yield palindromic cubes are palindromic themselves, as can be seen with:

$$11^3 = 1,331$$
$$101^3 = 1,030,301$$
$$1001^3 = 1,003,003,001$$
$$10001^3 = 1,000,300,030,001$$
$$100001^3 = 1,000,030,000,300,001$$

**Equation 13**

... and this continues on.

## THE NUMBER 197

The number 197 is a twin prime number with its partner 199.

The prime number 197 is the sum of all the two-digit prime numbers, which is:

197 = 11 + 13 + 17 + 19 + 23 + 29 + 31 + 37 + 41 + 43 + 47 + 53 + 59 + 61 + 67 + 71 + 73 + 79 + 83 + 89 + 97. The prime number 197 can also be obtained by taking the sum of the digits of all the two-digit prime numbers: 197 = 1 + 1 + 1 + 3 + 1 + 7 + 1 + 9 + 2 + 3 + 2 + 9 + 3 + 1 + 3 + 7 + 4 + 1 + 4 + 3 + 4 + 7 + 5 + 3 + 5 + 9 + 6 + 1 + 6 + 7 + 7 + 1 + 7 + 3 + 7 + 9 + 8 + 3 + 8 + 9 + 9 + 7.

The prime number 197 is also the sum of the first 12 prime numbers, which is:

197 = 2 + 3 + 5 + 7 + 11 + 13 + 17 + 19 + 23 + 29 + 31 + 37. Furthermore, the prime number 197 can also be expressed as the sum of 7 consecutive prime numbers as follows: 197 = 17 + 19 + 23 + 29 + 31 + 37 + 41.

The number 197 can also be exhibited in a rather curious way with repdigit numbers as follows: 197 = 111 + 9 + 77.

The number 197 can also be expressed in this comical fashion:
197 = (19 × 7) + (1 × 9 × 7) + 1. Notice the digit order.

## THE NUMBER 199

The number 199 is a 46th prime number and can be expressed as a sum of three consecutive primes: 199 = 61 + 67 + 71, or as the sum of five consecutive prime numbers: 199 = 31 + 37 + 41 + 43 + 47.

When the digits of the number 199 are permuted, each number results in a prime number as 991, and 919, which are both prime numbers.

When the number 199 is attached to the next prime number, 211, to get 199,211, or in the other direction 211,199, both resulting numbers are prime numbers.

## THE NUMBER 201

Although the number 201 is a composite number, the expression
$201^{14} + 2 = 175,689,094,255,155,802,434,244,515,642,803$, is a prime number,

and $201^{15} + 1 = 35,313,507,945,286,316,289,283,147,644,203,003$, is a prime number, and $201^{16} + 1 = 7,098,015,097,002,549,574,145,912,676,484,803,203$, is also a prime number.

## THE NUMBER 203

The composite number 203 can generate a prime number in a rather curious fashion (notice the exponents): $203^2 + 203^0 + 203^3 = 8,406,637$, which is a prime number.

Another surprising way in which the number 203 can generate a prime number is as follows:

$203^0 + 203^1 + 203^2 + 03^3 + 203^4 + 203^5 + 203^6 = 70,326,806,362,093$, which is a prime number.

## THE NUMBER 204

The square of the number 204 can be expressed as a sum of 3 consecutive cubes, as we show here: $204^2 = 41,616 = 23^3 + 24^3 + 25^3$.

However, $204^2 + 1 = 41,617$, which is a prime number.

## THE NUMBER 206

It should be well known that there are 206 bones in the adult human body.

The number 206 enables us to reach a prime number in a rather curious fashion:

$206^0 + 206^1 + 206^2 + 206^3 + 206^4 + 206^5 + 206^6 = 76,792,124,280,187$, which is a prime number.

The number 206 can be used to generate a twin pair of primes in the following way:

$206 + 206^2 + 1 = 42,643$, and $206 + 206^2 - 1 = 42,641$, which are consecutive primes, thus making them twin primes.

## THE NUMBER 208

Once again, we have a happy number the number 208 because $2^2 + 0^2 + 8^2 = 68$, and $6^2 + 8^2 = 100$, which then leads to 1, making 208 a happy number.

The number 208 can also be expressed as the sum of the squares of the first five prime numbers as shown: $208 = 2^2 + 3^2 + 5^2 + 7^2 + 11^2$.

## THE NUMBER 210

The number 210 is a triangular number (preceded by the triangular number 190), and it is also a pentagonal number (preceded by the pentagonal number 176), which is quite unusual. The smallest number that satisfies this condition is the number 1 and the next larger number that satisfies this condition is the number 40,755, which is 285th triangular number and the 165th pentagonal number. The *n*th triangular number is $T_n = \dfrac{n(n+1)}{2}$ and the *n*th pentagonal number is $P_n = \dfrac{n(3n-1)}{2}$.

The number 210 is equal to the product of the first four prime numbers ($210 = 2 \times 3 \times 5 \times 7$) and the sum of eight consecutive prime numbers $13 + 17 + 19 + 23 + 29 + 31 + 37 + 41 = 210$.

A curious aspect of the number 210 is that a series of prime numbers exists where the common difference is 210 as shown here: 47, 257, 467, 677, 887, 1,097, and 1,307.

## THE NUMBER 212

The boiling point of water is 212° Fahrenheit. The number 212 was the original telephone area code for New York City, and was selected because of the permissible options for an area code the number 212 required the least amount of dial movement. The first digit could not be a 1, and the second 2 digits could not be 11; therefore, 212 was the best option.

The number 212 in a clever way enables us to get a prime number as follows: $\dfrac{212}{2 \times 1 \times 2} = 53$, which is a prime number.

A prime number can also be generated by taking the number 212 to decreasing powers beginning with 10, as shown here:
$$212^{10} + 212^9 + 212^8 + 212^7 + 212^6 + 212^5 + 212^4 + 212^3 + 212^2 + 212^1 + 212^0 = 184,251,916,941,751,188,170,917,$$ which is, in fact, a prime number.

## THE NUMBER 216

Although the number 216 is a perfect cube, $6^3$, it is also the sum of its three preceding cubes:

$$216 = 3^3 + 4^3 + 5^3.$$

The number 216 can also generate three related prime numbers as follows:

- $6 \times 216 + 1 = 1,297$ is the 211th prime number;
- $12 \times 216 + 1 = 2,593$ is the 378th prime number, and
- $18 \times 216 + 1 = 3,889$, which is the 539th prime number.

Prime numbers can also generate the number 216 as follows: $\left(13^2 + 11^2\right) - \left(7^2 + 5^2\right) = 216$.

The number 216 is also the sum of a twin prime-number pair, as $107 + 109 = 216$.

A multiplicative magic square can also be created with a constant product of 216 as shown in Figure 69, where the product of each row, column, and diagonal is 216.

| | | |
|:---:|:---:|:---:|
| 3 | 4 | 18 |
| 36 | 6 | 1 |
| 3 | 4 | 18 |

Figure 69

## THE NUMBER 220

The voltage level in the European countries is 220 volts.

The number 220 is a sum of four consecutive prime numbers, namely, $47 + 53 + 59 + 61 = 220$.

With the number 220 we will introduce the concept of *amiable numbers*. Two numbers are considered *amicable*, if the sum of the proper divisors (divisors excluding the original number) of one number equals the second number, *and* the sum of the proper divisors of the second number equals the first number. The smallest pair of amicable numbers are: 220 and 284.

The divisors of **220** (other than 220 itself) are 1, 2, 4, 5, 10, 11, 20, 22, 44, 55, and 110.

Their sum is $1 + 2 + 4 + 5 + 10 + 11 + 20 + 22 + 44 + 55 + 110 = \mathbf{284}$.

The divisors of **284** (other than 284 itself) are 1, 2, 4, 71, and 142, and their sum is $1 + 2 + 4 + 71 + 142 = \mathbf{220}$. This shows that the two numbers are amicable numbers.

This pair of amicable numbers was already known to Pythagoras about 500 BCE.

A second pair of amicable numbers is 17,296 and 18,416. Its discovery is usually attributed to the French mathematician, Pierre de Fermat (1607–1665), although there is evidence that this discovery was anticipated by the Moroccan mathematician Ibn al-Banna al-Marrakushi al-Azdi (1256–c.1321).

The sum of the proper divisors of 17,296 is

$$1 + 2 + 4 + 8 + 16 + 23 + 46 + 47 + 92 + 94 + 184 + 188 + 368$$
$$+ \; 376 + 752 + 1,081 + 2,162 + 4,324 + 8,648 = 18,416.$$

The sum of the proper divisors of 18,416 is $1 + 2 + 4 + 8 + 16 + 1,151 + 2,302 + 4,604 + 9,208 = 17,296$.

Thus, they, too, are truly amicable numbers!

The French mathematician, René Descartes (1596–1650) found another pair of amicable numbers: 9,363,584, and 9,437,056. By 1747, the Swiss mathematician, Leonhard Euler (1707–1783), discovered 60 pairs of amicable numbers, yet he seemed to have overlooked the second smallest pair, namely, 1,184 and 1,210, which was discovered in 1866 by the 16-year-old B. Nicolò I. Paganini (1850–?): The sum of the factors of 1,184 is $1 + 2 + 4 + 8 + 16 + 32 + 37 + 74 + 148 + 296 + 592 = 1,210$.

And the sum of the factors of 1,210 is $1 + 2 + 5 + 10 + 11 + 22 + 55 + 110 + 121 + 242 + 605 = 1,184$.

To date more than 363,000 pairs of amicable numbers have identified, yet we do not know if there are an infinite number of such pairs. The following table provides a list of the first 108 amicable numbers. An ambitious reader might want to verify the "friendliness" of each of these pairs. Going beyond this list, we will eventually stumble on an even larger pair of amicable numbers: 111,448,537,712 and 118,853,793,424.

Readers who wish to pursue a search for additional amicable numbers might want to use the following method for finding amicable numbers: Consider the numbers $a = 3 \times 2^n - 1$, $b = 3 \times 2^{n-1} - 1$, and $c = 3^2 \times 2^{2n-1} - 1$, where $n$ is an integer $\geq 2$.

If $a$, $b$, and $c$ are prime numbers, then $2^n \times a \times b$ and $2^n \times c$ are amicable numbers. (For $n \leq 200$, only $n = 2, 4$, and 7 would give us $a$, $b$, and $c$ to be prime numbers.)

| | First Number | Second Number | Year of discovery |
|---|---|---|---|
| | | Table of Amicable Numbers | |
| 1 | 220 | 284 | ca. 500 BCE- |
| 2 | 1184 | 1210 | 1860 |
| 3 | 2620 | 2924 | 1747 |
| 4 | 5020 | 5564 | 1747 |
| 5 | 6232 | 6368 | 1747 |
| 6 | 10744 | 10856 | 1747 |
| 7 | 12285 | 14595 | 1939 |
| 8 | 17296 | 18416 | ca. 1310/1636 |
| 9 | 63020 | 76084 | 1747 |
| 10 | 66928 | 66992 | 1747 |
| 11 | 67095 | 71145 | 1747 |
| 12 | 69615 | 87633 | 1747 |
| 13 | 79750 | 88730 | 1964 |
| 14 | 100485 | 124155 | 1747 |
| 15 | 122265 | 139815 | 1747 |
| 16 | 122368 | 123152 | 1941/42 |
| 17 | 141664 | 153176 | 1747 |
| 18 | 142310 | 168730 | 1747 |
| 19 | 171856 | 176336 | 1747 |
| 20 | 176272 | 180848 | 1747 |
| 21 | 185368 | 203432 | 1966 |
| 22 | 196724 | 202444 | 1747 |
| 23 | 280540 | 365084 | 1966 |
| 24 | 308620 | 389924 | 1747 |
| 25 | 319550 | 430402 | 1966 |
| 26 | 356408 | 399592 | 1921 |
| 27 | 437456 | 455344 | 1747 |
| 28 | 469028 | 486178 | 1966 |
| 29 | 503056 | 514736 | 1747 |
| 30 | 522405 | 525915 | 1747 |
| 31 | 600392 | 669688 | 1921 |
| 32 | 609928 | 686072 | 1747 |
| 33 | 624184 | 691256 | 1921 |
| 34 | 635624 | 712216 | 1921 |
| 35 | 643336 | 652664 | 1747 |
| 36 | 667964 | 783556 | 1966 |
| 37 | 726104 | 796696 | 1921 |
| 38 | 802725 | 863835 | 1966 |
| 39 | 879712 | 901424 | 1966 |
| 40 | 898216 | 980984 | 1747 |
| 41 | 947835 | 1125765 | 1946 |
| 42 | 998104 | 1043096 | 1966 |
| 43 | 1077890 | 1099390 | 1966 |
| 44 | 1154450 | 1189150 | 1957 |
| 45 | 1156870 | 1292570 | 1946 |
| 46 | 1175265 | 1438983 | 1747 |
| 47 | 1185376 | 1286744 | 1929 |
| 48 | 1280565 | 1340235 | 1747 |
| 49 | 1328470 | 1483850 | 1966 |
| 50 | 1358595 | 1486845 | 1747 |
| 51 | 1392368 | 1464592 | 1747 |
| 52 | 1466150 | 1747930 | 1966 |

*(Continued)*

| 53 | 1468324 | 1749212 | 1967 |
|-----|---------|----------|------|
| 54 | 1511930 | 1598470 | 1946 |
| 55 | 1669910 | 2062570 | 1966 |
| 56 | 1798875 | 1870245 | 1967 |
| 57 | 2082464 | 2090656 | 1747 |
| 58 | 2236570 | 2429030 | 1966 |
| 59 | 2652728 | 2941672 | 1921 |
| 60 | 2723792 | 2874064 | 1929 |
| 61 | 2728726 | 3077354 | 1966 |
| 62 | 2739704 | 2928136 | 1747 |
| 63 | 2802416 | 2947216 | 1747 |
| 64 | 2803580 | 3716164 | 1967 |
| 65 | 3276856 | 3721544 | 1747 |
| 66 | 3606850 | 3892670 | 1967 |
| 67 | 3786904 | 4300136 | 1747 |
| 68 | 3805264 | 4006736 | 1929 |
| 69 | 4238984 | 4314616 | 1967 |
| 70 | 4246130 | 4488910 | 1747 |
| 71 | 4259750 | 4445050 | 1966 |
| 72 | 4482765 | 5120595 | 1957 |
| 73 | 4532710 | 6135962 | 1957 |
| 74 | 4604776 | 5162744 | 1966 |
| 75 | 5123090 | 5504110 | 1966 |
| 76 | 5147032 | 5843048 | 1747 |
| 77 | 5232010 | 5799542 | 1967 |
| 78 | 5357625 | 5684679 | 1966 |
| 79 | 5385310 | 5812130 | 1967 |
| 80 | 5459176 | 5495264 | 1967 |
| 81 | 5726072 | 6369928 | 1921 |
| 82 | 5730615 | 6088905 | 1966 |
| 83 | 5864660 | 7489324 | 1967 |
| 84 | 6329416 | 6371384 | 1966 |
| 85 | 6377175 | 6680025 | 1966 |
| 86 | 6955216 | 7418864 | 1946 |
| 87 | 6993610 | 7158710 | 1957 |
| 88 | 7275532 | 7471508 | 1967 |
| 89 | 7288930 | 8221598 | 1966 |
| 90 | 7489112 | 7674088 | 1966 |
| 91 | 7577350 | 8493050 | 1966 |
| 92 | 7677248 | 7684672 | 1884 |
| 93 | 7800544 | 7916696 | 1929 |
| 94 | 7850512 | 8052488 | 1966 |
| 95 | 8262136 | 8369864 | 1966 |
| 96 | 8619765 | 9627915 | 1957 |
| 97 | 8666860 | 10638356 | 1966 |
| 98 | 8754130 | 10893230 | 1946 |
| 99 | 8826070 | 10043690 | 1967 |
| 100 | 9071685 | 9498555 | 1946 |
| 101 | 9199496 | 9592504 | 1929 |
| 102 | 9206925 | 10791795 | 1967 |
| 103 | 9339704 | 9892936 | 1966 |
| 104 | 9363584 | 9437056 | ca. 1600/1638 |
| 105 | 9478910 | 11049730 | 1967 |
| 106 | 9491625 | 10950615 | 1967 |
| 107 | 9660950 | 10025290 | 1966 |
| 108 | 9773505 | 11791935 | 1967 |

Figure 70

Inspecting the list of amicable numbers in the preceding table, we notice that each pair is either a pair of odd numbers or a pair of even numbers. To date, we do not know if there is a pair of amicable numbers, where one is odd and one is even. We also do not know if any pair of amicable numbers is relatively prime (i.e., the numbers have no common factor other than 1). Perhaps these open questions contribute to our continued fascination with amicable numbers.

## THE NUMBER 227

The number 227 is the 49th prime number. If any digit is deleted from this three-digit number, the result will be a composite number.

The number 227 can be used to generate a prime number in a rather curious fashion: $2^{227} - 227^2 = 215,679,573,337,205,118,357,336,120,696,157,$ $045,389,097,155,380,324,579,848,828,881,942,199$, which is a prime number.

One can also use the first four prime numbers to generate the number 227 as follows:

$$227 = (2 \times 3 \times 5 \times 7) + 2 + 3 + 5 + 7.$$

The number 227 is the only three-digit prime number, where the product of its digits is a perfect number, namely, 28.

## THE NUMBER 229

The number 229 is the second half of a twin pair of prime numbers along with the number 227.

The number 229 can be expressed as $4^4 - 3^3 = 229$.

When the number 229 is added to its reversal 922, the resulting number, 1,151 is the 190th prime number.

There are times when strange manipulations create prime numbers. This can be done with the number 229. Suppose we replace each digit with its square, to get the number 4,481, which is the 608th prime number. Now we will replace each digit by its cube to get the number 88,729, which is the 8,593rd prime number. If we now take the difference of these two numbers, $88,729 - 4,481 = 84,248$, which, although it is not a prime number, it is a palindromic number. When we take the sum of these two numbers along with the number 229, we get $4,481 + 88,729 + 229$, which is another palindromic number 93,439.

When the number 2 is taken to the 229th power, the result has a very unusual characteristic. The 69-digit number has only one 0 amongst its digits: $2^{229}$ = 862,718,293,348,820,473,429,344,482,784,628,181,556,388,621,521, 298,319,395,315,527,974,912.

## THE NUMBER 231

The reciprocal of 231 is the smallest fraction in the sum of unit fractions which is equal to 1 as follows: $1 = \dfrac{1}{3} + \dfrac{1}{5} + \dfrac{1}{7} + \dfrac{1}{9} + \dfrac{1}{11} + \dfrac{1}{15} + \dfrac{1}{35} + \dfrac{1}{45} + \dfrac{1}{231}$.

## THE NUMBER 232

The number 232 is the first of three consecutive numbers, where each can be expressed as the sum of two squares.: $32 = 6^2 + 14^2$, $233 = 8^2 + 13^2$, $234 = 3^2 + 15^2$. It should be noted that there is no sequence of three consecutive numbers prior to these three where this relationship would hold true.

## THE NUMBER 233

The number 233 is a prime number and is the only known Fibonacci prime number with the sum of its digits is also Fibonacci number: $2 + 3 + 3 = 8$, which is the sixth Fibonacci number.

Using the number 233, we can generate a prime number as follows: $233 + (2 + 3 + 3) = 241$.

The number 233, which is composed of prime digits, can be expressed as a sum of 11 (a prime number) consecutive prime numbers: $233 = 5 + 7 + 11 + 13 + 17 + 19 + 23 + 29 + 31 + 37 + 41$.

## THE NUMBER 239

The smallest factor of the palindromic number 1,234,567,654,321 is the number 239 because this number is equal to $239^2 \times 4,649^2$. And even more surprising is that the product of $239 \times 4,649 = 1,111,111$.

The number 239 can also be represented as $\sqrt{2 \times 169^2 - 1}$.

The 239th prime is the number 1,499 which is a prime number. The reversal of 1,499 is 9,941, and it, too, is a prime number.

There are 239 prime numbers less than 1,500.

## THE NUMBER 240

The number 240 has more divisors than any previous number. These 20 divisors are: 1, 2, 3, 4, 5, 6, 8, 10, 12, 15, 16, 20, 24, 30, 40, 48, 60, 80, 120, and 240.

The number 240 can be expressed as a sum of consecutive prime numbers, either as: $240 = 53 + 59 + 61 + 67$, or as $240 = 17 + 19 + 23 + 29 + 31 + 37 + 41 + 43$.

## THE NUMBER 241

The number 241 is a prime number and the twin prime partner of 239.

When the number 241 is added to its reversal 142, the result is a palindromic number 383. Then seen from another point of view of reversals, the number 241 can be expressed as the sum of a number and its reversal: $241 = 170 + 071$.

The number 241 can be set equal to a configuration using only the digit 2, as follows: $\dfrac{22^2 - 2}{2} = 241$.

## THE NUMBER 243

The number 243 is a fifth power number as it can be expressed as $3^5$.

The number 243 is also equal to the sum of the first three palindromic prime numbers, namely, $343 = 11 + 101 + 131$.

## THE NUMBER 250

The number 250 can be expressed as a sum of two cubes: $5^3 + 5^3$.

The number 250 can also be expressed as a sum of two squares in two different ways: $250 = 13^2 + 9^2$, and $250 = 15^2 + 5^2$.

## THE NUMBER 251

The number 251 is a prime number and can be expressed as sum of three cubes in two different ways: $251 = 1^3 + 3^3 + 3^3$, and $251 = 2^3 + 3^3 + 6^3$.

The number 251 can also be expressed as a sum of three consecutive prime numbers ($251 = 79 + 83 + 89$) as well as the sum of seven consecutive prime numbers ($251 = 23 + 29 + 31 + 37 + 41 + 43 + 47$).

The number can be shown equal to a combination of prime numbers as $251 = (13 \times 17) + (1 + 17)$, which are all prime numbers.

The number 251 can be expressed in an unusual fashion by using only the first three prime numbers in the following way: $250 = 2^3 + 3^5$.

Looking back at the number 127, the number 251 shares the same de Polignac problem in that it cannot be expressed as a sum of a power of 2 and a prime number.

### THE NUMBER 252

The number 252 is divisible by the sum of its digits because $\dfrac{252}{2 + 5 + 2} = 28$, which is a perfect number.

The number 252 can be expressed as a sum of the cubes of its divisors, namely, $252 = 1^3 + 2^3 + 3^3 + 6^3$, and as the sum of products of prime number cubes: $252 = (1^3 + 2^3) \times (1^3 + 3^3)$.

### THE NUMBER 256

The number 256 is equal to the eighth power of 2, namely, $2^8$.

The famous Hungarian mathematician Paul Erdös (1913–1996) conjectured that the highest power of 2 that can be expressed as a sum of distinct powers of 3 is the number 256. In other words, $256 = 3^5 + 3^2 + 3^1 + 3^0$.

### THE NUMBER 257

The number 257 is smallest three-digit prime number with no two digits the same.

The number $257 = 4^4 + 1$, also $2^8 + 1$. Furthermore, the number 257 can also be expressed as $2^{2^3} + 1$.

The number 257 is comprised of only prime digits and it can be equal to the sum of three prime numbers, each of which is comprised of only prime digits, namely, $3 + 27 + 227 = 257$.

## THE NUMBER 263

The prime number 263 is equidistant from the previous prime number, 257, and the successor prime number 269, which means that it is the arithmetic mean of these two primes: 257 and 269.

The number 263 is 1 greater than twice a prime number, namely, $(2 \times 131) + 1 = 263$. This is often referred to as a Sophie Germain (1776–1831) prime number, named after the French mathematician.

The number 263 is the smallest prime number that is the sum of primes whose beginning and ending digits are a 1. That is, $263 = 11 + 101 + 151$. Note also, that each of these four numbers have a digit sum that is a prime number.

The number 263 is the only three-digit prime number where the product of the units digit and hundreds digit is equal to the tens digit.

## THE NUMBER 264

The number 264 when squared yields a very unusual palindromic number, $264^2 = 69,696$, which is also a palindrome when flipped upside down, 96,969.

The number $264^2$ is equal to the sum of the twin pair of prime numbers: $34,847 + 34,849$.

The number 264 can also be arrived at by raising the digits to consecutive prime numbers in reverse: $264 = 2^5 + 6^3 + 4^2$.

## THE NUMBER 271

The number 271 is a twin prime number with 269, and strangely enough, a factor of the repdigit number 11,111 because $271 \times 41 = 11,111$.

The reciprocal of 271 also has an unusual five-place repetition as we can see here: $\dfrac{1}{271} = 0.003690036900369....$

When the number 271 is taken to the second power and to the third power each is reversed to become prime numbers. Consider $271^2 = 73,441$ and then its reversal 14,437 is a prime number. Then considering the number $271^3 = 19,902,511$, we find that its reversal 11,250,991 is also a prime number.

## THE NUMBER 274

The number 274 lends itself to a very symmetric pattern for generating a prime number as we can see here where the number 274's digits appear as exponents and subtrahends as well as the number itself: $(274 - 2)^2 + (274 - 7)^7 + (274 - 4)^4 = 96,733,934,731,005,907$, which is a prime number.

## THE NUMBER 281

The number 281 is a prime number and is the sum of the consecutive primes up to 43:

$$2 + 3 + 5 + 7 + 11 + 13 + 17 + 19 + 23 + 29 + 31 + 37 + 41 + 43 = 281.$$

Were we to add the consecutive primes up to the number 281, the result would also be the prime number 7,699.

We can also get the number 281 in a rather symmetric (palindromic) array of arithmetic as follows:

$$281 = (9 \times 8) + (7 \times 6) + (5 \times 4) + (3 \times 2) + 1 + (2 \times 3) + (4 \times 5) + (6 \times 7) + (8 \times 9).$$

The number 281 is the first of three consecutive numbers that generate the same result (6) when their digits are continuously multiplied as shown:

- For the number 281: $2 \times 8 \times 1 = 16$, and $1 \times 6 = 6$.
- For the number 282: $2 \times 8 \times 2 = 32$, and $3 \times 2 = 6$.
- For the number 283: $2 \times 8 \times 3 = 48$, and $4 \times 8 = 32$, and $3 \times 2 = 6$.

## THE NUMBER 283

The distance from the East to West borders of Pennsylvania is 283 miles.
The number 283 is a twin prime with the number 281.
The number 283 is $(2 \times 8 \times 3) + (2 + 8 + 3) = 61$st prime number.
The number 283 allows itself to be expressed rather attractively as follows:

$$283 = \frac{(6! - 5! - 4! - 3! - 2! - 1! - 0!)}{2} = \frac{566}{2} = 283.$$

## THE NUMBER 284

Recall our discussion of the number 220, which is an amicable partner with 284. We offer here to give the number 284 equal treatment and to show that they are truly amicable.

The number 220 is a sum of four consecutive prime numbers, namely, $220 = 47 + 53 + 59 + 61$.

The number 220 will reintroduce the concept of *amiable numbers*. Two numbers are considered *amicable*, if the sum of the proper divisors (divisors excluding the original number) of one number equals the second number, *and* the sum of the proper divisors of the second number equals the first number. The smallest pair of amicable numbers are 220 and 284.

The divisors of **220** (other than 220 itself) are $1, 2, 4, 5, 10, 11, 20, 22, 44,$ 55, and 110.

Their sum is $1 + 2 + 4 + 5 + 10 + 11 + 20 + 22 + 44 + 55 + 110 = \mathbf{284}$.

The divisors of **284** (other than 284 itself) are $1, 2, 4, 71,$ and $142,$ and their sum is $1 + 2 + 4 + 71 + 142 = \mathbf{220}$. This shows that the two numbers are amicable numbers.

## THE NUMBER 286

A cute method to get a prime number can be generated by the number 286 as follows:

$286^{(2 + 8 + 6)} + (2 + 8 + 6) - 1 = 2,003,809,955,918,727,361,975,035,$ $+ (6 \times 7) + (8 \times 9)$, which is a prime number.

## THE NUMBER 287

Although the number 287 is not a prime number it can be obtained from three different sums of consecutive prime numbers as follows:

- $287 = 89 + 97 + 101$;
- $287 = 47 + 53 + 59 + 61 + 67$;
- $287 = 17 + 19 + 23 + 29 + 31 + 37 + 41 + 43 + 47$.

## THE NUMBER 288

The number 288 is the smallest number, which is a multiple of nine and is composed entirely of even digits. A motivated reader might try to find a smaller number that has those characteristics and would be frustrated at the inability to find one.

The number 288 can also be expressed as the product of the first four factorials: $288 = 1! \times 2! \times 3! \times 4!$.

## THE NUMBER 293

The number 293 is a prime number. There are 293 ways in which to break up a $1 amount using the following coins: penny, nickel, dime, quarter, half-dollar, and dollar.

The number 293 can be expressed as the sum of three special prime numbers, 11, 101, and 181, which are the first three prime numbers that can be flipped over without changing their value.

## THE NUMBER 295

Although the number 295 is not a prime number, there is a very neat way of generating a prime number using the number 295 and its digits in order: $(2 \times 295) + (9 \times 295) + (5 \times 295) + 1 = 4,721$, which is a prime number.

## THE NUMBER 296

Although the number 296 is not a prime number, there are some unusual way to generate prime numbers using the number 295 as follows:

- $296 \times 2! \times 9! \times 6! + 1 = 154,673,971,201$, which is a prime number.
- $2! \times 9! \times 6! + (2 + 9 + 6) = 522,547,217$, which is a prime number.
- $2! \times 9! \times 6! - (2 + 9 + 6) = 522,547,183$, which is a prime number.

## THE NUMBER 297

The number 297 is the fifth Kaprekar number. To demonstrate that taking the square of 297, we get $297^2 = 88,209$. Split it into two parts, and strangely enough, the sum of the two numbers of the split results in the original number: $88 + 209 = 297$. Such a number is called a *Kaprekar number*, named after the Indian mathematician Dattaraya Ramchandra Kaprekar (1905–1986) who discovered such numbers. (For more about Kaprekar numbers see page 208.)

If we take a cyclic permutation of the number 297, say 729 and repeat this process: $729^2 = 531,441$, and then split the number to get the numbers 531 and 441, and amazingly, we find that their sum is 972, a number with which we started.

The number 297 also provides a neat arrangement to generate a prime number as follows:

$(2 \times 297) + (9 \times 297) + (7 \times 297) + 1 = 5,347$, which is a prime number.


## THE NUMBER 307

The number 307 is the 63rd prime number and the square of 307 is a palindromic number: $307^2 = 94,249$.

The reverse of number 307, namely, 703 is a Kaprekar number (see page 208) because $703^2 = 494,209$, and $494 + 209 = 703$.

The number 307 can be written as the sum of specific prime numbers, where all the digits are represented: $307 = 2 + 5 + 41 + 67 + 83 + 109$.


## THE NUMBER 360

In baseball, when a player hits a homerun, he must run 360 feet rounding the bases. Perhaps the number 360 is most famous for the fact that there are 360° in one complete revolution.

The number 360 has 24 divisors as follows: 2, 2, 3, 4, 5, 6, 8, 9, 10, 12, 15, 18, 20, 24, 30, 36, 40, 45, 60, 72, 90, 120, 180, and 360. Furthermore, it is the smallest number divisible by every natural number from 1 to 10 except the number 7.

The number 360 is the sum of two twin prime numbers: $360 = 179 + 181$.


## THE NUMBER 362

The number 362 can be interestingly expressed in two ways:

$3! + 6! + 2! - 1 = 727$, and $3! \times 6! \times 2! + 1 = 8,641$. Both of these numbers are prime numbers.


## THE NUMBER 365

The number 365 represents the number of days in a non-leap year.

The number 365 can be expressed as the sum of consecutive squares as: $365 = 10^2 + 11^2 + 12^2$ and $365 = 13^2 + 14^2$. Notice that the squares listed are all consecutive numbers. This pattern continues as can also be seen with the following sums: $21^2 + 22^2 + 23^2 + 24^2 = 25^2 + 26^2 + 27^2$.

## THE NUMBER 371

The number 371 is equal to the sum of the cues of its digits: $371 = 3^3 + 7^3 + 1^3$. The number 371 is equal to the sum of the consecutive primes from 7 to 53, as shown here:

$$7 + 11 + 13 + 17 + 19 + 23 + 29 + 31 + 37 + 41 + 43 + 47 + 53 = 371.$$

## THE NUMBER 373

Water boils at 373 on the Kelvin scale. (To be more exact, 373.15 K.)
The number 373 is a palindromic prime number comprised of prime digits.
The number 373 can be expressed as the sum of the squares of five consecutive primes as shown here: $373 = 3^2 + 5^2 + 7^2 + 11^2 + 13^2$.
One can also arrive at the number 373 with the initial consecutive numbers in order: $373 = (12 \times 34) - 5 - 6 - 7 - 8 - 9$.

## THE NUMBER 376

The number 376 shares a property with the number 625 in that its powers all end in the same three digits from which they started. We can see that in the following list:

| | | |
|---|---|---:|
| $376^1$ | $=$ | **376** |
| $376^2$ | $=$ | 141,**376** |
| $376^3$ | $=$ | 53,157,**376** |
| $376^4$ | $=$ | 19,987,173,**376** |
| $376^5$ | $=$ | 7,515,177,189,**376** |
| $376^6$ | $=$ | 2,825,706,623,205,**376** |
| $376^7$ | $=$ | 1,062,465,690,325,221,**376** |
| $376^8$ | $=$ | 399,487,099,562,283,237,**376** |
| $376^9$ | $=$ | 150,207,149,435,418,497,253,**376** |
| $376^{10}$ | $=$ | 56,477,888,187,717,354,967,269,**376** |

**Equation 15**

## THE NUMBERS 399 AND 400

The number 399 has a peculiar property, that is, 399 is divisible by 3, 7, and 19, while the next number 399 + 1 is divisible by 3 + 1, 7 + 1, and 19 + 1. Arithmetically seen: $\frac{399}{3} = 133$, $\frac{399}{7} = 57$, $\frac{399}{19} = 21$, while $\frac{400}{4} = 100$, $\frac{400}{8} = 50$, $\frac{400}{20} = 20$.

This is a highly unusual situation!

## THE NUMBER 499

The number 499 is a prime number.

The number 499 = 497 + 2 and the reverse of 499, namely 994 = 497 × 2. Another curiosity!

## THE NUMBER 504

The number 504 lends itself to an unusual relationship: 24 × 21 = 504 = 42 × 12.

## THE NUMBER 518

The number 518 can be expressed as the sum of the consecutive powers of its digits as shown here: $518 = 5^1 + 1^2 + 8^3$.

## THE NUMBER 598

The number 598 is yet another number that can be expressed as the sum of the consecutive powers of its digits, as we can see here: $598 = 5^1 + 9^2 + 8^3$.

## THE NUMBER 625

The number 625 is equal to $5^4$.

The number 625 can also be expressed as $625 = 2^4 + 2^4 + 3^4 + 4^4 + 4^4$.

When the number 625 is squared, the result is 390,625. When the number 625 is cubed we get 244,140,625. Taking this a step further, $625^4 = 152{,}587{,}890{,}625$, and $625^5 = 95{,}367{,}431{,}640{,}625$. What should be noticed here is that every power of 625 ends in the same three digits, namely, 625. Here are just a few examples of this:

$$625^k$$

| | | |
|---|---|---:|
| $625^1$ | $=$ | **625** |
| $625^2$ | $=$ | 390,**625** |
| $625^3$ | $=$ | 244,140,**625** |
| $625^4$ | $=$ | 152,587,890,**625** |
| $625^5$ | $=$ | 95,367,431,640,**625** |
| $625^6$ | $=$ | 59,604,644,775,390,**625** |
| $625^7$ | $=$ | 37,252,902,984,619,140,**625** |
| $625^8$ | $=$ | 23,283,064,365,386,962,890,**625** |
| $625^9$ | $=$ | 14,551,915,228,366,851,806,640,**625** |
| $625^{10}$ | $=$ | 9,094,947,017,729,282,379,150,390,**625** |

**Equation 16**

## THE NUMBER 651

The number 651 introduces an unusual product relationship with its reversal, namely, $651 \times 156 = 101{,}556 = 372 \times 273$. Notice how both products are reversals of one another—highly unusual!

## THE NUMBER 659

The number 659 is the first of a twin pair of primes, which precedes the next pair of twin primes, namely 809 and 811, which is one of the largest known spans between twin primes of 148 numbers.

## THE NUMBER 666

The number 666 is often referred to as "the number of the beast" which could likely be from the Book of Revelation 13:15–18, which reads "And that no man might buy or sell, save he that had the mark, or the name of the beast, or the number of his name. Here is wisdom. Let him that hath understanding count the number of the beast: for it is the number of a man; and his number is six hundred three-score and six."

However, in mathematics it has some interesting aspects. The number 666 is the sum of the squares of the first seven prime numbers, that is, $666 = 2^3 + 3^2 + 5^2 + 7^2 + 11^2 + 13^2 + 17^2$. And also the number 666 is the sum of the first 36 natural numbers: $1 + 2 + 3 + \dots + 34 + 35 + 36 = 666$.

The sum of the digits of 666 ($6 + 6 + 6 = 18$) is equal to the sum of the digits of its prime factors. Its prime factors are: $666 = 2 \times 3 \times 3 \times 37$, and the sum of the digits of the prime factors:

$$2 + 3 + 3 + 3 + 7 = 18 = 6 + 6 + 6.$$

The value of $\pi$ to 144 places is $\pi \approx 3.14159265358979323846264338327950288419716939937510582097494459230781640628620899862803482534211706798214808651328230664709384460955058223172535 9 \dots$ The sum of these 144 digits is 666.

Two consecutive triangular numbers, 15 and 21 also play a role with the number 666, in that $15^2 + 21^2 = 225 + 441 = 666$.

The number 666, in a rather creative way, can create a prime number as follows: $666 \times 66 \times 6 + 1 = 263,737$, which is a prime number.

The first gap of 666 numbers between two pairs of twin prime numbers occurs between the pair of twin prime numbers, 774,131 and 774,133, and the next pair of twin prime numbers 774797 and 774,799.

We can get a pair of twin prime numbers as follows $666 + 666 + 666 + 1 = 1,999$, and $666 + 666 + 666 - 1 = 1,997$.

Here are a few delectable number relationships, which lead to 666:

$$666 = 1^6 - 2^6 + 3^6$$
$$666 = (6 + 6 + 6) + (6^3 + 6^3 + 6^3)$$
$$666 = (6^4 - 6^4 + 6^4) - (6^3 + 6^3 + 6^3) + (6 + 6 + 6)$$
$$666 = 5^3 + 6^3 + 7^3 - (6 + 6 + 6)$$
$$666 = (2^1 \times 3^2) + (2^3 \times 3^4)$$

We can even generate 666 by representing each of its three digits in terms of 1, 2, and 3:

$$6 = 1 + 2 + 3$$
$$6 = 1 \times 2 \times 3$$
$$6 = \sqrt{1^3 + 2^3 + 3^3}$$

Therefore, $666 = \big[(100) \times (1 + 2 + 3)\big] + \big[(10) \times (1 \times 2 \times 3)\big]$
$$+\left[\sqrt{1^3 + 2^3 + 3^3}\right]$$

Incidentally, when written in Roman numeral form, the number 666 uses all the Roman symbols prior to the M (1,000) in descending order: 666 = DCLXVI.

## THE NUMBER 691

The number 691 is a prime number composed of digits none of which is a prime number.

When the number 691 is flipped over and reversed to get 169, a perfect square number results: $169 = 13^2$.

When the number 691 is reversed to get 196, another perfect square number results: $196 = 14^2$.

## THE NUMBER 701

The number 701 is the lead number of the following prime numbers: 701, 7,001, 70,001, and 700,001. This cannot be extended to inserting five 0s because $7,000,001 = 197 \times 35,533$.

The prime number $701 = 1^0 + 2^1 + 3^2 + 4^3 + 5^4$.

The square of the number 701, that is, $701^2 = 491,401$, which is a number consisting of only square digits.

## THE NUMBERS 714–715—RUTH–AARON NUMBERS

For many years the goal for most home-run-hitting baseball players was to reach, or surpass, the longtime record for career home runs set by Babe Ruth at 714 home runs. On April 8, 1974, the Atlanta Braves' slugger Hank Aaron hit his 715th career home run.[29] This prompted the American mathematician Carl B. Pomerance (1944–) to popularize the notion (through the suggestion of one of his students) that these two numbers 714 and 715 are related by dint of the fact that they are two consecutive numbers whose prime factor sums are equal:

$$714 = 2 \times 3 \times 7 \times 17, \text{ and } 2 + 3 + 7 + 17 = 29,$$
$$715 = 5 \times 11 \times 13, \text{ and } 5 + 11 + 13 = 29.$$

We can extend the list of Ruth–Aaron numbers, if we consider only numbers with distinct prime factors. In this case we have the following pairs or Ruth–Aaron numbers:

(5, 6), (24, 25), (49, 50), (77, 78), (104, 105), (153, 154), (369, 370), (492, 493), (714, 715), (1682, 1683), (2107, 2108).

If multiplicities are not counted (so that a factor of $2^3$ counts only a single 2), we get the following pair:

(24, 25): 24 $= 2 \times 2 \times 2 \times 3$, and $2 + 3 = 5$,
25 $= 5 \times 5$, and $5 = 5$.

If we consider repeating prime factors, then the following numbers would also qualify as Ruth–Aaron numbers:

$$\left(5, 6\right), \left(8, 9\right), \left(15, 16\right), \left(77, 78\right), \left(125, 126\right), \left(714, 715\right), \left(948, 949\right),$$
$$\left(1,330, 1,331\right).$$

(8, 9): 8 $= 2 \times 2 \times 2$, and $2 + 2 + 2 = 6$,
9 $= 3 \times 3$,    and $3 + 3 = 6$.

There are also pairs of unusual Ruth–Aaron numbers, where the sums of all the factors are equal, and where the sums of the factors, without repetition of each of the factors, are equal. One such pair is: (7,129,199 and 7,129,200), where $7,129,199 = 7 \times 11^2 \times 19 \times 443$, and $7,129,200 = 2^4 \times 3 \times 5^2 \times 13 \times 457$.

Without repeating the factors of each of the numbers:

$$7 + 11 + 19 + 443 = 480 = 2 + 3 + 5 + 13 + 457,$$

The sums taking into account the repetition of the factors:

$$7 + 11 + 11 + 19 + 443 = 491 = 2 + 2 + 2 + 2 + 3 + 5 + 5 + 13 + 457.$$

We can extend the Ruth–Aaron pairs to consider Ruth–Aaron triplets, that is, three consecutive numbers, where the sums of the factors are equal. Here is one such triplet using the three numbers 89,460,295, 89,460,295 and 89,460,296:

$$89,460,294 = 2 \times 3 \times 7 \times 11 \times 23 \times 8,419$$
$$89,460,295 = 5 \times 4,201 \times 4,259$$
$$89,460,296 = 2 \times 2 \times 2 \times 31 \times 43 \times 8,389$$

The sums of the factors are as follows:

$$2 + 3 + 7 + 11 + 23 + 8,419 = 5 + 4,201 + 4,259$$
$$= 2 + 31 + 43 + 8,389 = 8,465.$$

This is the first such example without repetition!

Another such Ruth–Aaron triplet is the following: 151,165,960,539, 151,165,960,540, and 151,165,960,541.

$$151,165,960,539 = 3 \times 11 \times 11 \times 83 \times 2,081 \times 2,411$$
$$151,165,960,540 = 2 \times 2 \times 5 \times 7 \times 293 \times 1,193 \times 3,089$$
$$151,165,960,541 = 23 \times 29 \times 157 \times 359 \times 4,021$$

The sums of the factors are:

$$3 + 11 + 83 + 2,081 + 2,411 = 2 + 5 + 7 + 293 + 1,193 + 3,089$$
$$= 23 + 29 + 157 + 359 + 4,021 = 4,589.$$

Here we have, yet, another Ruth–Aaron triplet, but with repeating the factors: 417,162, 417,163, and 417,164.

$$417,162 = 2 \times 3 \times 251 \times 277$$
$$417,163 = 17 \times 53 \times 463$$
$$417,164 = 2 \times 2 \times 11 \times 19 \times 499$$

The sums of the prime factors are:

$$2 + 3 + 251 + 277 = 17 + 53 + 463 = 2 + 2 + 11 + 19 + 499 = 533$$

The last Ruth–Aaron triplet so far to have been found is: 6,913,943,284, 6,913,943,285, and 6,913,943,286.

$$6,913,943,284 = 2 \times 2 \times 37 \times 89 \times 101 \times 5,197$$
$$6,913,943,285 = 5 \times 283 \times 1,259 \times 3,881$$
$$6,913,943,286 = 2 \times 3 \times 167 \times 2,549 \times 2,707$$

The sums of the prime factors are:

$$2 + 2 + 37 + 89 + 101 + 5,197 = 5 + 283 + 1,259 + 3,881$$
$$= 2 + 3 + 167 + 2,549 + 2,707 = 5,428.$$

Who would have thought that the two home-run kings would make such a contribution to mathematics? By the way, if we remind ourselves that this discussion began with 714 and 715, and we take the sum of these numbers 714 + 715 = 1,429, which is a prime number, as is its reversal 9,241. Furthermore, other arrangements of the digits of this number are also prime numbers: 4,219, 4,129, 9,412, 1,249.

Interestingly, Pomerance found that up to the number 20,000 there are 26 pairs of Ruth–Aaron numbers, the largest of which is 18,490 and 18,491. One of the most famous mathematicians of the 20th century, Paul Erdös (1913–1996), proved that there are an infinite number of Ruth–Aaron numbers. Feel free to search for others!

## THE NUMBER 720

The number 720 is equal to 6!

The number 720 can also be written as $2 \times 3 \times 4 \times 5 \times 6 = 720 = 8 \times 9 \times 10$.

The uniqueness of the number 720 is that it can be expressed as the quotient of two factorials: $720 = 6! = \dfrac{10!}{7!}$

The number 720 can also be written as the sum of 2 squares: $720 = 24^2 + 12^2$.

## THE NUMBER 727

The number 727 can be expressed as $727 + 1! + (1 + 2)! + (1 + 2 + 3)!$

The number 727 can be expressed as the sum of a number and its reversal in three different ways: $116 + 611 = 727, 215 + 512 = 727$, and $314 + 413 = 727$.

## THE NUMBER 729

The number $729 = 9^3 = 27^2 = 1^3 + 3^3 + 4^3 + 5^3 + 8^3$.

## THE NUMBER 733

The number 733 is the 130th prime number.

The number $733 = 1^7 + 2^6 + 3^5 + 4^4 + 5^3 + 6^2 + 7^1 + 8^0$. Notice the pattern of the exponents!

The number $733 = 7 + 3! + (3!)!$

We can also get the number 733 by adding 6! to the sixth prime number, 13, so that $733 = 6! + 13$.

## THE NUMBER 757

The number 757 is the 134th prime number.

The number 757 is the largest three-digit palindromic prime number that is not the sum of consecutive composite numbers.

The number 757 is the sum of seven consecutive prime numbers: $757 = 97 + 101 + 103 + 107 + 109 + 113 + 127$.

An interesting feature of the number 757 is that the hundreds digit can be replaced with a power of 2 and still be a prime number: 157, 257, 457, and 857.

## THE NUMBERS 780 AND 990

The number 780 along with the number 990 are a pair of triangular numbers, the 39th and 44th, respectively, whose sum and difference are also triangular numbers: $990 - 780 = 210$, is the 20th triangular number, and $990 + 780 = 1,770$ is the 59th triangular number. This is a highly unusual situation!

## THE NUMBERS 788–793

The numbers 788, 789, 790, 791, 792, and 793 are divisible by the consecutive prime numbers: 2, 3, 5, 7, 11, and 13, respectively. Rather curious!

## THE NUMBER 823

The number 823 is a prime number with a unique feature:

$$823 = \frac{12,345}{1 + 2 + 3 + 4 + 5}$$

## THE NUMBER 836

The number 836, when it is squared, results in a rather unusual number, namely, 698,896, which is not only a palindrome but also can be flipped upside down to become the palindrome $968,869 = 3 \times 32,323$, and another palindrome appears. This time, the palindrome 32,323 is also a prime number.

## THE NUMBER 839

The number 839 is a prime number that can generate another prime number in a most unusual way: $(839 - 8)^8 + (839 - 3)^3 + (839 - 9)^9 = 187,167,$ $664,551,255,812,641,144,897$, which is a prime number. Notice the exponents.

## THE NUMBER 863

The number 863 is a prime number, which can be written as the sum of 15 consecutive prime numbers as: 29 + 31 + 37 + 41 + 43 + 47 + 53 + 59 + 61 + 67 + 71 + 73 + 79 + 83 + 89 = 863.

A prime number can be created by taking the sum of each of the three digits of the number 863 repeated the number of times determined by the digit number. That is,

88,888,888 + 666,666 + 333 = 89,555,887, which is a prime number.

## THE NUMBER 873

The number 873 = 1! + 2! + 3! + 4! + 5! + 6!

## THE NUMBER 891

The number 891 can be partitioned into five groups of three numbers each, where each group has the same sum and the same product as shown here:

- 891 = 6 + 480 + 495, and 6 × 480 × 495 = 1,425,600
- 891 = 11 + 160 + 810, and 11 × 160 × 810 = 1,425,600
- 891 = 12 + 144 + 825, and 12 × 144 × 825 = 1,425,600
- 891 = 20 + 81 + 880, and 20 × 81 × 880 = 1,425,600
- 891 = 33 + 48 + 900, and 33 × 48 × 900 = 1,425,600

## THE NUMBER 907

Beginning with the number 907, which is a prime number, a series of further prime numbers can be constructed with additional 9s as follows:

907, 9,907, 99,907, 999,907, 9,999,907, are each prime numbers. Please note that with six 9s the number 99,999,907 = 7 × 19 × 751,879, obviously not a prime number!

## THE NUMBER 911

The number 911 has taken on a variety of meanings in the United States. For many years it was, and still is, the universal telephone number to contact the police in an emergency situation. On September 11, 2001, it took on another unfortunate meaning as it reminds people of the devastating attack on the World Trade Center in New York and on the Pentagon, and the plane crash in Pennsylvania. On a more positive note, on September 11, 1609, the English explorer Henry Hudson (1565–1611) sailed into New York Harbor for the first time, essentially initiating the beginning of New York City.

In Europe, the term 9/11 refers to November 9, which also recalls an historic event. On November 9, 1989, the Berlin wall was breached for the first time, indicating an end to the German Democratic Republic and unifying Germany. On a rather negative note, November 9, 1938, was the night that the German Nazis attacked Jewish persons and property in Germany and Austria, demolishing stores and synagogues, and is recalled under the name Kristallnacht, meaning "night of broken glass." On November 9, 1965, the East Coast of the United States and Canada experienced one of the biggest blackouts when electric power shut down.

In mathematics, the number 911 is 156th prime number.

The number 911 is the first of nine consecutive prime numbers, which have a very curious characteristic. If one deletes the hundreds digit 9 from each of the following nine consecutive prime numbers, the result will be a list of prime numbers as well. The nine consecutive prime numbers are: 911, 919, 929, 937, 941, 947, 953, 967, and 971. Deleting the hundreds digit from each of these prime numbers leaves us with the following prime numbers: 11, 19, 29, 37, 41, 47, 53, 67, and 71.

## THE NUMBER 919

The number 919 is a prime number that pairs interestingly with the number 1,459 as follows:

$919 = 1^3 + 4^3 + 5^3 + 9^3$ and notice the pattern of the breakdown of the number 1,459 equals $9^3 + 1^3 + 9^3$.

Not only is the number 919 a palindromic number but the next prime number, 929, is also a palindromic number.

It is believed that the number 919 is the smallest palindromic prime number, which is equal to the difference of two consecutive cubes: $919 = 18^3 - 17^3$.

A cute relationship that the number 919 offers is the following:

$$919 = 91 + 81 + 71 + 61 + 51 + 41 + 31 + 21 + 11 + 1 + 11 + 21 + 31 +$$
$$41 + 51 + 61 + 71 + 81 + 91.$$

Note the palindromic relationship.

## THE NUMBER 929

The number 929, which is the twin prime number partner of 919, can be expressed as $9^{29} + 92^9 = 4,710,128,697,718,406,198,208,160,361$, which is also a prime number. Again, notice the pattern of the exponents as compared to the actual bases.

The number 929 can also be expressed symmetrically as $(1!)^5 + (2!)^4 + (3!)^3 + (4!)^2 + (5!)^1 = 929$. Appreciate the symmetry in this relationship.

When the number 929 is added to the product of its digits, $929 + (9 \times 2 \times 9) = 1091$, which is a prime number as is its reversal 1901, also a prime number.

## THE NUMBER 1,001

Because the number 1,001 is the product of the three prime numbers 7, 11, and 13, a clever algorithm can be made to determine if a very large number is divisible by the number 7, 11, and 13. This is best explained in the context of a specific situation. Consider the number 23,592,840,010,274,535. To determine if it is divisible by 7, 11, and 13, we partition it into groups of three digits, and take the sum of the 2nd, 4th, and 6th groups of three digits: $274 + 840 + 23 = 1,137$. We then take the sum of the remaining groups of 3 digits: $535 + 010 + 592 = 1,137$ and subtract the two sums: $1137 - 1137 = 0$. If the resulting difference is divisible by, 7, 11, and 13 (i.e., by 1,001) then the original number is divisible by 1,001. Here we had a difference of 0. Therefore, the original number is divisible by 7, 11, and 13 because 0 is divisible by any number.

## THE NUMBER 1,009

The number 1,009 is a prime number as is its reversal 9,001.

The number 1,009 can be expressed as the sum of three cubes in two different ways: $1,009 = 1^3 + 2^3 + 10^3$, and $1,009 = 4^3 + 6^3 + 9^3$.

The number 109 can be expressed as the sum of a prime number and its reversal in two different ways: $1,009 = 257 + 752$, and $1,009 = 653 + 356$.

## THE NUMBER 1,031

The number 1,031 is a prime number, as is its reversal 1,301.

Furthermore, other arrangements of the four digits of the number 1,031 are also prime numbers, such as 1,103, 1,013, and 3,011.

The number 1,031 can be expressed using consecutive prime numbers as follows: $(2 \times 3 \times 5) + (7 \times 11 \times 13)$.

## THE NUMBER 1,033

The number 1,033 is the twin prime number partner of 1,031 and can be expressed using the digits $1 - 5$ as follows: $1,033 = 1 + 2^3 + 4^5$.

The number 1,033 can also be expressed in this unusual way: $1,033 = 8^1 + 8^0 + 8^3 + 8^3$, where the number 1,033 is exhibited amongst the exponents.

## THE NUMBER 1,089

The number 1,089 has a number of oddities attached to it. One characteristic of this number can be seen by taking the reciprocal of this number (1,089) and getting the following: $\dfrac{1}{1,089} = .\overline{0009182736455463728191}$. With the exception of the first three zeros and the last 1, we have a palindromic number: 918,273,645,546,372,819 because it reads the same in both directions.

Furthermore, if we multiply $1,089 \times 5$, we also get a palindromic number 5,445, and if we multiply 1,089 by 9, we get 9,801 the reverse of the original number. By the way, the only other number of four or fewer digits, whose multiple is the reverse of the original number is 2,178 because $2,178 \times 4 = 8,712$.

The number $1,089 = 33^2 = 65^2 - 56^2$. This is the only two-digit illustration of this pattern.

Let us now do multiplication by 9 of some numbers, which are modifications of 1,089, say 10,989, 109,989, 1,099,989, 10,999,989, and so on, and then marvel at the results.

$$10989 \cdot 9 = 98901$$

$$109989 \cdot 9 = 989901$$

$$1099989 \cdot 9 = 9899901$$

$$10999989 \cdot 9 = 98999901$$

**Equation 17**

... and so on.

Returning to the number 1,089 we find that it has embedded in it a very entertaining oddity. Suppose you select any three-digit number whose units digit and hundreds digits are not the same, and then reverse that number. Now subtract the two numbers you have (obviously, the larger minus the smaller). Once again reverse the digits of this arrived at difference and add this new number to the difference. The result will always be 1,089.

To see that this works, we will choose any randomly selected three-digit number, say 732. We now subtract $732 - 237 = 495$. Reversing the digits of 495 we get 594, and now we add these last two numbers: $495 + 594 = 1,089$. Yes, this will hold true for all such three-digit numbers—amazing! This is a cute little "trick" that can be justified with simple algebra.

Here is the justification using elementary algebra. We shall represent the arbitrarily selected three-digit number, $\overline{htu}$ as $100h + 10t + u$, where $h$ represents the hundreds digit, $t$ represents the tens digit, and $u$ represents the units digit. The number with the digits reversed is then $100u + 10t + h$. We will let $h > u$, which would be the case either in the number you selected or its reverse. In the subtraction, $u - h < 0$; therefore, take 1 from the tens place (of the minuend), making the units place $10 + u$. Because the tens digits of the two numbers to be subtracted are equal, and 1 was taken from the tens digit of the minuend, then the value of this digit is $10(t - 1)$. The hundreds digit of the minuend is $h - 1$ because 1 was taken away to enable subtraction in the tens place, making the value of the tens digit $10(t - 1) + 100 = 10(t + 9)$.

When we do the first subtraction, we get:

$$
\begin{array}{lll}
100(h-1) & +10(t+9)+(u+10) & \\
100u & +10t & +h \\
\hline
100(h-u-1) & +10(9) & +u-h+10
\end{array}
$$

**Equation 18**

Reversing the digits of this difference $100(h - u - 1) + 10 \times 9 + (u - h + 10)$ gives us:

$100(u - h + 10) + 10 \times 9 + (h - u - 1)$. By adding these last two expressions, gives us:

$100(h - u - 1) + 10 \times 9 + (u - h + 10) + 100(u - h + 10) + 10 \times 9 + (h - u - 1) = 1,000 + 90 - 1 = \mathbf{1,089}$. This algebraic justification enables us to inspect the general case of this arithmetic process, allowing us to guarantee that this process holds true for all numbers.

## THE NUMBER 1,100

The number 1,100 can be expressed as the sum of two palindromic prime numbers in three different ways as shown here: $1,100 = 181 + 919$;  $1,100 = 313 + 787$; $1,100 = 373 + 727$.

## THE NUMBER 1,111

The repunit number 1,111 is the sum of the digits of the first 100 prime numbers.

The number 111 can also be expressed as the sum of the squares of six consecutive numbers as shown here: $1,111 = 11^2 + 12^2 + 13^2 + 14^2 + 15^2 + 16^2$.

With special consideration of the units digits, we find that the number $1,111 = 56^2 - 45^2$, where previously we found that $11 = 6^2 - 5^2$. This pattern continues as we show on page 210.

## THE NUMBER 1,141

The number 1,141 enables its sixth power to be expressed as the sum of seven other sixth-power numbers as shown: $1,141^6 = 74^6 + 234^6 + 402^6 + 474^6 + 702^6 + 894^6 + 1,077^6 = 2,206,550,475,483,180,841$.

## THE NUMBER 1,233

The number 1,233 is one of those unusual numbers that can be broken into the following pattern: $1,233 = 12^2 + 33^2$.

## THE NUMBERS 1,309, 1,310, 1,311

The first three consecutive numbers that can each be expressed as the product of three distinct prime numbers are the numbers 1,309, 1,310, and 1,311 as shown here:

$1,309 = 7 \times 11 \times 17$, $1,310 = 2 \times 5 \times 131$, and $1,311 = 3 \times 19 \times 23$.

## THE NUMBER 1,331

The number 1,331 is a palindromic square, which equals $11^3$, and can be also expressed as a sum of the squares of three consecutive prime numbers as: $1,331 = 19^2 + 21^2 + 23^2$.

The number 1,331 is the smallest cube $(11^3)$ that can be expressed as a sum of three consecutive prime numbers, namely, 439, 443, and 449.

## THE NUMBER 1,361

The number 1,361 is a prime number but allows the creation of other prime numbers in a rather unusual way. By splitting up the 13 and the 61 and placing a **63** between them, one gets the number 136,**3**61, which is a prime number. This can be repeated by inserting 2 numbers 63 between the 13 and the 61 to get the number 13,**636,3**61, which is also a prime number. This can be repeated twice more to get two additional prime numbers, namely: 1,**363,636,3**61 and 136,**363,636,3**61. When another 63 is inserted, the result is a composite number $13,\mathbf{636,363,636,3}61 = 7 \times 1,321 \times 5,693 \times 10,661$.

## THE NUMBERS 1,375, 1,376, AND 1,377

The number 1,375, 1,376 and 1,377 is an unusual triple of consecutive numbers, each of which is divisible by a cubed prime number, which shows as follows: $1,375 = 5^3 \times 11$; $1,376 = 2^3 \times 172$; and $1,377 = 3^3 \times 51$.

## THE NUMBER 1,399

The number 1,399 is a prime number that can generate further primes by adding consecutive powers of 10 to it. Therefore, the following numbers are all prime numbers:

$1,399 + 10 = 1,409$

$1,399 + 100 = 1,499$

$1,399 + 1000 = 2,399$

$1,399 + 10,000 = 11,399$

$1,399 + 100,000 = 101,399$

This cannot be continued any further.

## THE NUMBER 1,427

There are times when one can create a rather curious arrangement of the digits of a prime number that yields another prime number. Such is the case with the prime number 1,427. Consider each term together with its powers and notice what you see: $142^7 + 14^{27} + 1^{427} = 8,819,763,977,946,282,294,620,364,$ $692,353$, which is a prime number.

## THE NUMBER 1,453

The number 1,453 is the largest prime number that can be represented in two different ways as a sum of a three-digit number and its reversal: $1,453 = 479 + 974$, and $1,453 = 677 + 776$.

## THE NUMBERS 1,458 AND 1,729

The numbers 1,458 and 1,729 share a curious pattern:

For 1,458, we have, $1 + 4 + 5 + 8 = 18$, then multiplied by its reversal: $18 \times 81 = 1,458$.

For 1,729, we have, $1 + 7 + 2 + 9 = 19$, then multiplied by its reversal: $19 \times 91 = 1,729$.

N.B. There is much more on the now-famous number 1,729 on page 185.

## THE NUMBER 1,634

The number 1,634 has a peculiar property that is shared with only two other four-digit numbers, namely, that the number can be equal to the sum of the fourth power of its digits, as we show here that $1,634 = 1^4 + 6^4 + 3^4 + 4^4$. The other two numbers sharing this characteristic are the numbers 8,208 and 9,474.

## THE NUMBER 1,681

At first glance the number 1,681 highlights the fact that is comprised of two square numbers, 16 and 81. However, what makes this number special is that aside from multiples of 100, this is the only four-digit number comprised of squares as we have here and is itself a perfect square, since $1,681 = 41^2$.

## THE NUMBER 1,760

The number 1,760 yards is equal to one mile.

## THE NUMBER 1,729

The number 1,729 became famous through an incident reported by the British mathematician G. H. Hardy (1877–1947) when he visited the brilliant Indian mathematician Srinivasa Ramanujan (1887–1920), who was lying ill in a hospital. To make small talk as he approached the ill Ramanujan, he mentioned that he had been riding in a taxicab no. 1729 and couldn't make much out of that number. Without hesitation, Ramanujan told him that 1,729 is a very interesting number, and it is the smallest number that can be expressed as a sum of two cubes in two different ways as: $1,729 = 12^3 + 1^3 = 10^3 + 9^3$. This completely amazed Hardy, who then wrote about this amazing experience. It is also the closing scene of a popular film "The Man Who Knew Infinity," based on a book (of same name as the film, 1991) by Robert Kanigel.

We might well give credit for this original finding to the famous Swiss mathematician Leonhard Euler (1707–1783), who stated that the smallest natural number that can be expressed as the sum of two cubes of natural numbers in two ways is 1,729.

That is, $1^3 + 12^3 = 1 + 1,728 = 1,729$, and $9^3 + 10^3 = 729 + 1,000 = 1,729$.

To date, the largest known number that can be expressed as a sum of two cubes in two different ways is: $885,623,890,831 = 7,511^3 + 7,730^3 = 8,759^3 + 5,978^3$.

To be consistent, the number $885,623,890,831$ can also be shown equal to the product of three prime numbers as follows: $885,623,890,831 = 3,943 \times 14,737 \times 15,241$.

Curiously, the number 1,729 can also be expressed as the product of three prime numbers: $1,729 = 7 \times 13 \times 19$.

Even more on the number $1,729 = 19 \times 91$, which can be expressed as 19, which is the sum of the digits of 1,729 times its reversal.

## THE NUMBERS 1,885, 1,886, AND 1,887

The first three consecutive numbers that can each be expressed as the product of three distinct prime numbers are the numbers 1,885, 1,886, and 1,887 as shown here:

$1,885 = 5 \times 13 \times 29$; $1,886 = 2 \times 23 \times 41$; and $1,887 = 3 \times 17 \times 37$.

## THE NUMBER 1,980

When from 1,980 its reverse number, 0,891, is subtracted the special number 1,089 results. (For more special features of the number 1,089, see page 180.)

## THE NUMBER 2,178

If we take the sum of the fourth powers of the digits of 2,178, we begin a scheme, which will get us back to the original number, 2,178, as follows: $2^4 + 1^4 + 7^4 + 8^4 = 6,514$, and now we take the fourth powers of 6,514: $6^4 + 5^4 + 1^4 + 4^4 = 2,178$, which was a number with which we began.

When the number 2,178 is multiplied by 4 its reversal appears: $2,178 \times 4 = 8,712$. If this is not enough of an amazement, then consider placing a 9 in the middle of the number 2,178 to get 21,978. Once again, we repeat the process: $21,978 \times 4 = 87,912$, which is its reversal. Surprisingly, this can be continued by placing 99 in the middle of 2,178 to get 219,978, and when multiplied by 4, we get $219,978 \times 4 = 879,912$, once again the reversal of the number with which we started. This can then be continued by placing 999 in the middle of 2,178 to get its reversal. And so on.

## THE NUMBER 2,418

We will use the number 2,418 to demonstrate a clever little arithmetic trick. This trick only holds for numbers where two adjacent digits do not have a sum exceeding 9. When we multiply $2,418 \times 11 = 26,598$, and then multiply its reversal by 11: $8,142 \times 11 = 89,562$, which is the reverse of the previous product we obtained.

## THE NUMBER 3,024

The number 3,024 can be expressed as the product of two numbers and a second time with the multiplying numbers with the digits reversed: $36 \times 84 = 3,024 = 63 \times 48$. (See "Special Number Characteristics," page 223.)

## THE NUMBER 3,435

The number 3,435 seems to be the only multidigit number that has the following property:

$$3,435 = 3^3 + 4^4 + 3^3 + 5^5.$$

## THE NUMBER 4,104

The number 4,104 can be expressed as the sum of two cubes in three different ways:

$$4,104 = 16^3 + 2^3 = 15^3 + 9^3 = (-12)^3 + 18^3.$$

## THE NUMBER 4,150

It is believed that the number 4,150 is the smallest number, which can be expressed as a sum of the fifth power of its digits as follows: $4,150 = 4^5 + 1^5 + 5^5 + 0^5$.

## THE NUMBER 4,356

The number 4,356 has an unusual characteristic. If it is divided by its reversal the result is:

$\dfrac{4356}{6534} = \dfrac{2}{3}$, or another way of stating it is that the number 3,456 is two-thirds of its reversal, 6,534.

## THE NUMBER 4,900

The number 4,900 can be expressed as a sum of the squares of the counting numbers from 1 to 24, as shown here: $4,900 = 1^2 + 2^2 + 3^2 + 4^2 + 5^2 + \ldots + 22^2 + 23^2 + 24^2 = 70^2$.

## THE NUMBER 5,213

The number 5,213 can be expressed as the sum of factorials from 1 through 7:

$$5,213 = 1! + 2! + 3! + 4! + 5! + 6! + 7!$$

## THE NUMBER 5,280

The number of feet in 1 mile is 5,280 feet.

The number 5,280 finds itself sandwiched between two twin prime numbers, namely, 5,279 and 5,281.

## THE NUMBER 5,965

The number, 5965 can follow a similar path as that shown previously for 6,205 as shown here:

$5,965 + 77^2 + 06^2$ and $59^2 + 65^2 = 7,706$. Again, the same relationship exhibited with the number, 6,205.

## THE NUMBER 6,174

The number 6,174 is known as the *Kaprekar constant*. This constant arises when one takes any four-digit number, and forms the largest and the smallest number from these digits, and then subtracts these two newly formed numbers. Continuing this process will always eventually result in the number 6,174. When the number 6,174 is arrived at, we continue the process of creating the largest and the smallest number, and then taking their difference $(7,641 - 1,467 = 6,174)$, we notice that we get back to 6,174. This is, therefore, called the *Kaprekar constant*. To demonstrate this with an example, we will carry out this process with the number 2,303 as follows:

- The largest number formed with these digits is: 3,320.
- The smallest number formed with these digits is: 0,233.
- The difference is: 3,087.
- The largest number formed with these digits is: 8,730.
- The smallest number formed with these digits is: 0,378.
- The difference is: 8,352.
- The largest number formed with these digits is: 8,532.
- The smallest number formed with these digits is: 2358.
- The difference is: 6,174.
- The largest number formed with these digits is: 7,641.
- The smallest number formed with these digits is: 1,467.
- The difference is: 6,174.

And so, the loop is formed, since you continue to get the number 6,174. Remember, all this began with an arbitrarily selected number, and will always end up with the number 6,174, which then gets you into an endless loop (i.e., continuously getting back to 6,174).

Another curious property of 6,174 is that it is divisible by the sum of its digits: $\dfrac{6174}{6+1+7+4} = \dfrac{6174}{18} = 343$.

## THE NUMBER 6,205

The number 6,205 lends itself to a rather unusual relationship, namely, that $6{,}205 = 38^2 + 69^2$, and in breaking up the number 6,205 into $62^2 + 05^2 = 3{,}869$. Admire the relationship!

## THE NUMBER 6,578

The number 6,578 is the smallest number that can be expressed as a sum of three fourth powers in two different ways: $6\ 578 = 1^4 + 2^4 + 9^4 = 3^4 + 7^4 + 8^4$.

## THE NUMBER 6,666

The square of the number 6,666 has a peculiar characteristic, where $6{,}666^2 = 44{,}435{,}556$, which when we separate into two parts and add them, we get: $4{,}443 + 5{,}556 = 9{,}999$.

This can also be done with the number $3{,}333^2$ and $7{,}777^2$, this is left is an entertainment for the reader!

(See repunits, page 209.)

## THE NUMBER 8,000

The number 8,000 is a cube, $20^3$, that can be expressed as a sum of four consecutive cubes: $8{,}000 = 11^3 + 12^3 + 13^3 + 14^3$.

## THE NUMBER 8,128

After the first three perfect numbers: 6, 28, 496, the number 8,128 is the fourth perfect number, as it is the sum of its proper divisors: $1 + 2 + 4 + 8 + 16 + 32 + 64 + 127 + 254 + 508 + 1{,}016 + 2{,}032 + 4{,}064 = 8{,}128$.

## THE NUMBER 8,191

The number 8,191 is a prime number and can be expressed in various interesting ways:

- $8,191 = 2^{13} - 1$.
- Also $8,191 = 1 + 90 + 90^2$, or more extensively as:
- $8,191 = 2^0 + 2^1 + 2^2 + 2^3 + 2^4 + 2^5 + 2^6 + 2^7 + 2^8 + 2^9 + 2^{10}$
  $+ 2^{11} + 2^{12}$.

## THE NUMBER 8,208

The number 8,208 is one of three four-digit numbers that can be expressed as a sum of the fourth power of its digits as we show here: $8,208 = 8^4 + 2^4 + 0^4 + 8^4$.

## THE NUMBER 9,474

The number 9,474 is one of three four-digit numbers that can be expressed as a sum of the fourth power of its digits as we show here: $9,474 = 9^4 + 4^4 + 7^4 + 4^4$.

## THE NUMBER 9,999

The number 9,999 has some unusual characteristics. Taking the square of 9,999 to get:

$9,999^2 = 99,980,001$, and then splitting that into two parts and finding the sum to be:

$9,998 + 0,001 = 9,999$, which is the number with which we began.

Now consider the cube of the number 9,999, which is $9,999^3 = 9,997 \times 0,002 \times 9,999$, $9,999^3 = 9,997 \times 0,002 \times 9,999$, and when we take the sum of those three equal-length parts, we get: $9,997 + 0,002 + 9,999 = 19,998$. Unfortunately, this cannot be extended to the fourth power.

## THE NUMBER 27,594

Although there is nothing special about this number, its factors can be arranged in the same order with the same product as follows: $27,594 = 73 \times 9 \times 42 = 7 \times 3,942$. Notice that the digits in the two products are in the same order.

## THE NUMBER 40,585

The number 40,585 can be shown to be equal to the sum of the factorials of its digits, as we show here: $40,585 = 4! + 0! + 5! + 8! + 5! = 24 + 1 + 120 + 40,320 + 120$.

## THE NUMBER 44,488

The number 44,488 is the first of five consecutive numbers each of which is a happy number; recall that happy numbers have the following characteristic, which involves squaring the digits of each number: $4^2 + 4^2 + 4^2 + 8^2 + 8^2 = 176$, and $1^2 + 7^2 + 6^2 = 86$, and $8^2 + 6^2 = 100$, then $1^2 + 0^2 + 0^2 = 1$ because we reached the number 1, we can, therefore, consider 44,488 a happy number.

## THE NUMBER 54,748

The number 54,748 can be expressed as a sum of the fifth power of its digits as shown here:

$$54,748 = 5^5 + 4^5 + 7^5 + 4^5 + 8^5.$$

## THE NUMBER 63,504

The number 63,504 can be written as a product of two numbers, one being the reverse of the other in two different ways as shown: $63,504 = 441 \times 144 = 252 \times 252$.

## THE NUMBER 69,696

The number 69,696 is a palindromic number and also a perfect square because $69,696 = 264^2$.

However, when flipped around the number becomes 96,969, which is also a palindrome with an interesting factorization: $96,969 = 3 \times 32,323$, where once again a palindrome appears as part of the product.

## THE NUMBER 90,625

The number 90,625 has a peculiar characteristic that when the number is taken to various powers its last five digits will be 90,625 as we can see with the following:

$90625^2 = 8212890625$

$90625^3 = 744293212890625$

$90625^4 = 67451572418212890625$

$90625^5 = 6112798750400543212890625$

$90625^7 = 50203747549676336348056793212890625$

$90625^8 = 4549714621689417981542646884918212890625$

$90625^9 = 412317887590603504577302373945713043212890625$

$90625^{10} = 37366308562898442602318027638830244541168212890625$

$90625^{11} = 3386321713512671360835071254768990911543369293212890625$

$90625^{12} = 306885405287085842075678332463439801358617842197418212890625$

**Equation 19**

This pattern then continues on.

## THE NUMBER 142,857

The number 142,857 is particularly attractive for its various unusual aspects. When the number 142,857 is multiplied by the numbers from 1 to 6, the result shows a cyclic permutation of the digits of the original number as we show in the following text:

$$142,857 \times 1 = 142,857$$
$$142,857 \times 2 = 285,714$$
$$142,857 \times 3 = 428,571$$
$$142,857 \times 4 = 571,428$$
$$142,857 \times 5 = 714,258$$
$$142,857 \times 6 = 857,142$$

When this number 142,857 is multiplied by 7, the product is 999,999. Multiplying the number 142,857 by larger numbers produces some other interesting results that will be worth the readers efforts to explore.

The number 142,857 can be partitioned into two parts and added: $142 + 857 = 999$. It should be noted that the number 999 is a factor of 142,857 because $142,857 \div 999 = 143$.

The nines can reappear by adding the pairs of digits of our number 142,857 as follows:

$$14 + 28 + 57 = 99.$$

This can be taken a giant step further if we take our number 142,857 and multiply it by any large number, say 243, and then split the number and marvel over the result as follows:

$142,857 \times 243 = 34,714,251$. Then partitioning the number in 3s from right to left: $34 + 714 + 251 = 999$. For a further example, $142,857 \times 3,636$, $= 519,428,052$ then partitioning and adding we get: $519 + 428 + 052 = 999$.

We know that the number 9 is a factor of 142,857 because the sum of the digits $1 + 4 + 2 + 8 + 5 + 7 = 27$, which is a multiple of 9, and therefore, the original number $142,857 \div 9 = 15,873$.

When we square the number 142,857 and split it another interesting result emerges.

$142,857^2 = 20,408,122,449$. Splitting this number and adding the two parts:

$204,081 + 22,449 = 142,857$, back to the original number.

The reciprocal of 142,857 also produces a surprise: $\dfrac{1}{142857} = 0.\overline{000007}$, the appearance of the 7 could let us expose an extra surprise as we can see from the reciprocal of 7: $\dfrac{1}{7} = 0.\overline{142857}$, and lo and behold, our original number 142,857 reappears in the repeating decimal.

Furthermore, the number 142,857 is divisible by both 11 and 111 because $142,857 = 111 \times 287 = 11 \times 12,987$.


## THE NUMBER 144,648


The number 144,648 is a number that can be written as the product of two reversals of one another in two different ways: $144,648 = 861 \times 168 = 492 \times 294$.

## THE NUMBER 175,560

The number 175,560 has the unexpected feature that it can be expressed as the product of consecutive numbers in two different ways: $175,560 = 55 \times 56 \times 57 = 19 \times 20 \times 21 \times 22$.

## THE NUMBER 183,184

The number 183,184 is a number composed of two consecutive three-digit numbers, which also happens to be a perfect square, as we can see where $183,184 = 428^2$. This is not the only number that has this unusual property. There are three other six-digit numbers that share the property of being composed of two consecutive three-digit numbers, which are perfect squares, and they are:

$$328,329 = 573^2, \ 528,529 = 727^2, \ 715,716 = 846^2.$$

## THE AMAZING NUMBER 193,939

We now present a truly remarkable prime number, that is, 193,939. Were we to write it in reverse, we would get 939,391, which is also a prime number. By considering other permutations of this number, we get prime numbers as shown here:

193 939
939 391
393 919
939 193
391 939
919 393

## THE NUMBER 396,733

The number 396,733 is the first of a pair of consecutive prime numbers that differs from the next prime number 396,833 by 100 numbers.

## THE NUMBER 490,689

The number 490,869 can be broken up into the sum of three cubes in two different ways and then provides a further surprise following that: $490,869 = 4^3 + 60^3 + 65^3 = 8^3 + 25^3 + 78^3$. Now comes a surprise using the same base numbers we find an incredible coincidence of equality: $4 \times 60 \times 65 = 8 \times 25 \times 78$.

## THE NUMBER 510,510

The number 510,510 is asymmetric number, as can easily be seen. However, the number can be expressed as the product of the first seven prime numbers, namely, $510,510 = 2 \times 3 \times 5 \times 7 \times 11 \times 13 \times 17$.

Coincidentally, the number 510,510 can also be expressed as a product of two consecutive numbers, $510,510 = 714 \times 715$, which are often referred to as the Ruth–Aaron numbers (see page 172).

## THE NUMBER 548,834

The number 548,834 has one peculiar aspect to it namely it can be expressed as the sum of six sixth powers of its digits: $548,834 = 5^6 + 4^6 + 8^6 + 8^6 + 3^6 + 4^6$.

## THE NUMBER 666,666

The number 666,666 will appear as the number of the devil used twice. However, it has another strange appearance, namely in geometry. A right triangle whose sides have lengths 693, 1,924, and 2,045 has an area equal to 666,666 square units.

## THE NUMBER 698,896

The number 698,896 is a palindrome clearly seen and also a square number because $698,896 = 836^2$. For more about this number see the number 836.

## THE NUMBER 739,397

The number 739,397 has a most unusual characteristic. If we remove one digit at a time from left to right, we always create a new prime number, as we can show here with the following prime numbers: 739,397, then 39,397, then 9,397, then 397, then 97, and finally 7. This can also be done in the opposite direction from right to left and once again generating a list of prime numbers: 739,397, then 73,939, then 7,393, then 739, then 73, and finally 7.

## THE NUMBER 798,644

The number 798,644 when it is squared results in a very unusual palindrome: 637,832,238,736.

## THE NUMBER 1,048,576

The number 1,048,576 is a perfect square because $1,048,576 = 1,024^2$. The curious thing about this number is that each of the six rearrangements of its digits also produces a perfect square each time as we can see from the following:

$$1,056,784 = 1,028^2$$
$$1,085,764 = 1,042^2$$
$$5,740,816 = 2,396^2$$
$$5,764,801 = 2,401^2$$
$$6,754,801 = 2,599$$
$$7,845,601 = 2,801^2$$

## THE NUMBER 1,741,725

The number 1,741,725 can be expressed as the sum of the seventh power of its digits as:

$$1,741,725 = 1^7 + 7^7 + 4^7 + 1^7 + 7^7 + 2^7 + 5^7.$$

## THE NUMBER 3,628,800

The number 3,628,800 does not appear to carry any special properties at first sight. However, $3,628,800 = 10!$ and furthermore, $10! = 6! \times 7! = 3! \times 5! \times 7!$

## THE NUMBER 5,761,455

What makes this number have a special place in mathematics is that it represents the number of prime numbers less than 100,000,000.

## THE NUMBER 24,678,050

The number 24,678,050 has the unique characteristic, namely, that this eight-digit number can be expressed as a sum of its eight digits each to the eighth power: $24,678,050 = 2^8 + 4^8 + 6^8 + 7^8 + 8^8 + 0^8 + 5^8 + 0^8$.

## THE NUMBER 42,549,416

The number 42,549,416 can be expressed as the sum of two cubes in four different ways:

$$42,549,416 = 348^3 + 74^3 = 282^3 + 272^3 = (-2662)^3 + 2664^3$$
$$= (-475)^3 + 531^3.$$

## THE NUMBER 87,539,319

The number 87,539,319 can be expressed as the sum of two cubes in three different ways as follows:

$$87,539,319 = 167^3 + 436^3 = 228^3 + 423^3 = 255^3 + 414^3.$$

## THE NUMBER 256,103,393

The number 256,103,393 can be expressed as the sum of three fourth powers in two different ways:

$256,103,393 = 22^4 + 93^4 + 116^4 = 29^4 + 66^4 + 124^4$, and furthermore, consider the relationship of the bases used: $22 \times 93 \times 116 = 29 \times 66 \times 124$. Quite amazing!

## THE NUMBER 438,579,088

The number 438,579,088 has a rather unusual characteristic, namely, that the sum of the digits taken to the power indicated by the digit yields the original number 438,579,088.

Admire this relationship: $438,579,088 + 1 = 4^4 + 3^3 + 8^8 + 5^5 + 7^7 + 9^9 + 0^0 + 8^8 + 8^8$.

## THE NUMBER 1,375,298,099

The number 1,375,298,099 can be shown to be equal to the sum of three fifth power numbers in two different ways: $1,375,298,099 = 24^5 + 28^5 + 67^5 = 3^5 + 54^5 + 62^5$.

## THE NUMBER 2,438,195,760

The number 2,438,195,760 is divisible by every number from 2 through 18. An ambitious reader may wish to try to verify this property by using rules for divisibility indicated earlier.

## THE NUMBER 6,661,661,161

The number 6,661,661,161 is a number comprised of two digits. The unique thing about this number is that it is a number, which is a perfect square comprised of only two digits. We have earlier identified such squares of only two digits such as the number 16, the number 121, and others of this kind. However, this is thought to be the largest such number found to date. It should be noted that it is quite likely that computers will come up with larger numbers that are perfect squares and at the same time comprised of only two digits. The number $6,661,661,161 = 81,619^2$.

## THE NUMBER 9,876,543,210

The number 9,876,543,210 is a number where all digits are represented exactly once. When we subtract from this number its reversal, 0,123,456,789, the answer is 9,753,086,421. Each of these three numbers is comprised of all the digits used exactly once.

## THE NUMBER 15,527,402,881

The number 15,527,402,881 is the smallest fourth power that is the sum of four fourth power numbers as shown here:

$$15,527,402,881 = 353^4 = 30 + 120^4 + 272^4 + 315^4.$$

## THE NUMBER 719,737,131,179

The number 719,737,131,179 is a number where every pair of digits in either direction is a prime number.

## NOTES

1. The quadratic formula for solving for $x$ in the general quadratic equation $ax^2 + bx + c = 0$ is $x = \dfrac{-b \pm \sqrt{b^2 - 4ac}}{2a}$.

2. There is reason to believe that the letter $\phi$ was used because it is the first letter of the name of the celebrated Greek sculptor, Phidias (ca. 490–430 BCE) [in Greek: (Pheidias) ΦΕΙΛΙΑΣ or Φειδίας], who produced the famous statue of Zeus in the Temple of Olympia and supervised the construction of the Parthenon in Athens, Greece. His frequent use of the Golden Ratio in this glorious building is likely the reason for this attribution. It must be said that there is no direct evidence that Phidias consciously used this ratio. The American mathematician Mark Barr (1871–1950) was the first using the letter $\phi$ in about 1909. See Theodore Andrea Cook, *The Curves of Life*. [New York, Dover Publications, 1979].

3. A *unit square* is a square with side length of 1 unit.

4. "Unique prime divisors" refers to divisors of a number that are prime numbers and not used more than once. For example, the number 105 is a number with unique prime divisors: 3, 5, and 7, while 315 is a number that does not have unique prime divisors: 3, 3, 5, and 7 because the prime divisor, 3, is repeated.

5. An irrational number is one that cannot be expressed as a fraction that has integers in its numerator and denominator.

6. The proof in 1767 by the German mathematician, Johann Heinrich Lambert (1728–1777), had a flaw in it.

7. A transcendental number is one that is not the root of a polynomial equation with rational coefficients. Another way of saying this is that it is a number that cannot be expressed as a combination of the four basic arithmetic operations and root extraction. In other words, it is a number that cannot be expressed algebraically. $\pi$ is such a number.

8. The term *transcendental number* was introduced by Euler.

9. Alfred S. Posamentier and Noam Gordon, "An Astounding Revelation on the History of $\pi$." *Mathematics Teacher* 77(1), (January 1984): 52.

10. The "molten sea" was a gigantic bronze vessel for ritual ablutions in the court of the First Temple (966–955 B.C.). It was supported on the backs of 12 bronze oxen (volume ≈ 45,000 liters).

11. A cubit is the length of a person's fingertip to his elbow.

12. In those days Vilna was in Poland, while today the town is named Vilnius and is in Lithuania.

13. Note well, this is *not* what π was later on to represent, namely, the circumference.

14. Kenneth Appel and Wolfgang Haken, "The Solution of the Four-Color-Map Problem." *Scientific American* 237(4), (1977): 108–21.

15. For more about this famous problem, see *The Mathematics of Everyday Life* by A. S. Posamentier and C. Spreitzer. Prometheus Books, 2018.

16. For more about the Reuleaux polygons, see *The Circle: A Mathematical Exploration beyond the Line,* by A. S. Posamentier and R. Geretschläger. Prometheus Books, 2016.

17. The orthocenter of a triangle is the point of intersection the three altitudes of the triangle.

18. *Recherches sur la determination d'une hyperbole équilatèau moyen de quartes conditions données* (Paris, 1820).

19. The multiplicand is the number that is multiplied by another number, the multiplier. In arithmetic, the multiplicand and the multiplier are interchangeable, depending on how the problem is stated, because the result is the same if the two are reversed, for example, 2 × 3 and 3 × 2. In arithmetic performed by computers, however, the multiplicand is different from the multiplier because computer multiplication is usually performed as addition. Therefore, 2 × 3 means "add 2 three times," whereas 3 × 2 means "add 3 two times."

20. For the interested reader, here is a brief discussion about why this rule works as it does.

    Consider the number $\overline{abcde}$, where $a, \dots, e \in \{0, 1, 2, 3, \dots, 9\}$ and $a \neq 0$, whose value can be expressed as

$$N = 10^4 a + 10^3 b + 10^2 c + 10d + e = (11 - 1)^4 a + (11 - 1)^3 b + (11 - 1)^2 c$$
$$+ (11 - 1)d + e = [11M + (-1)^4]a + [11M + (-1)^3]b + [11M + (-1)^2]c$$
$$+ [11 + (-1)]d + e = 11M[a + b + c + d] + a - b + c - d + e,$$

    which implies that divisibility by 11 of $N$ depends on the divisibility of:

    $a - b + c - d + e = (a + c + e) - (b + d)$, the difference of the sums

    of the alternate digits.

    Note: $11M$ denotes a multiple of 11.

21. See Alfred S. Posamentier and Ingmar Lehmann, *The (Fabulous) Fibonacci Numbers.* Afterword by Herbert Hauptman, Nobel Laureate. Prometheus Books, 2007, pp. 27, 97 ff.

22. "Solution to Problem E36." *American Mathematical Monthly* 40 (1933): 607.

23. A phobia for the number 13.

24. A *Mersenne number* is a number of the form $M_n = 2^n - 1$, where $n$ is an integer. A *Mersenne prime* is a prime number of the form $M_n = 2^n - 1$.

25. "Hydroxydeoxycorticosterones" and "hydroxydesoxycorticosterone" have 27 letters.

26. For more on this property see Posamentier and Lehmann, *The (Fabulous) Fibonacci Numbers* (Prometheus Books, 2007).

27. This triangle is formed by beginning on top with 1, then the second row has 1, 1, then the third row is obtained by placing 1s at the end and adding the two numbers in the second row $(1 + 1 = 2)$ to get the 2. The fourth row is obtained the same way. After the end 1s are placed, the 3s are gotten from the sum of the two numbers above (to the right and left), that is $1 + 2 = 3$, and $2 + 1 = 3$.

28. Those of us who have lived through 1991 and 2002 will be the last generation who will have lived through two palindromic years for more than the next 1,000 years (assuming the current level of longevity).

29. It should be noted that although Aarons went on to hit 755 lifetime home runs, the current record for most lifetime home runs is held by Barry Bonds at 762 home runs.

# 2

# Special Number Characteristics

## PERFECT NUMBERS

Although we have covered this topic at various points earlier, such as with the number 6 and the number 28, you may be wondering now what makes a number perfect? We provide here a discussion of these unusual numbers for the more motivated reader. Mathematicians define a number to be a *perfect number* when the sum of its factors (excluding the number itself) is equal to the number. For example, the smallest perfect number is 6 because the sum of its factors (of its proper positive divisors): $1 + 2 + 3 = 6$. The next larger perfect number is 28 because, again, the sum of its factors: $1 + 2 + 4 + 7 + 14 = 28$. Perfect numbers have fascinated mathematicians for centuries. The ancient Greeks knew of the first four perfect numbers and even Euclid (ca. 300 BCE) established a formula for generating these perfect numbers. These first two perfect numbers were regarded by ancient biblical scholars as perfect in that the biblical creation was accomplished in 6 days and the lunar month as 28 days. The next two perfect numbers (496 and 8,128) are attributed to Nicomachus (60 − 120 CE). Yet, it was not until the 18th century that the Swiss mathematician Leonhard Euler (1707–1783) proved that a formula developed by Euclid would generate all even perfect numbers.

Let us now consider what Euclid developed as a generalized formula for generating additional perfect numbers. The formula for generating even perfect numbers is $2^{n-1}(2^n - 1)$, with the condition that $2^n - 1$ is a prime number, and is referred to as a Mersenne prime number, $M_p{}^1$. For one thing, if $n$ is a composite number, then $2^n - 1$ will surely be composite, thereby making such a value of $n$ not one to generate a perfect number from Euclid's formula. We can demonstrate this with some very elementary algebra, as we will show here.

Suppose $n$ is an even composite number, say $2x$, the expression $2^n - 1$ then becomes the difference of two squares, which is always factorable, and is, therefore, a composite number. $2^{2x} - 1 = (2^x - 1)(2^x + 1)$. If $n$ is an odd composite number as in the expression $2^{pq} - 1$, it becomes a factorable term as $2^{pq} - 1 = (2^p)^q - 1 = (2^p - 1)\big((2^p)^{q-1} + (2^p)^{q-2} + (2^p)^{q-3} + \ldots + (2^2) + 1\big)$.

This is not to say that whenever $n$ is prime that $2^n - 1$ will also be prime. For example, when $n = 11$, we have $2^{11} - 1 = 2{,}048 - 1 = 2{,}047 = 23 \times 89$, and is therefore, not prime. Thus, we have to be careful to make sure that $n$ is a prime number and that it also generates $2^n - 1$ to be a prime.

There have been some bumps in the road that mathematicians encountered as they pursued the search for further perfect numbers. For example, the French mathematician Marin Mersenne (1588–1648), who studied perfect numbers, corrected the list of 24 perfect numbers, published by the 17th-century mathematician Peter Bungus in his book *Numerorun Mysteria* (1644), stating that only eight of these were correct (i.e., the first eight on the list in the table in Figure 70). However, Mersenne offered to add three more numbers to this list of perfect numbers (namely, those where $n$ in Euclid's formula had values: 67, 127, and 257). By 1947, it was proved that only the number 127 was correct, and at that time two more perfect numbers with values of $n$ being 89 and 107 were added to the list of perfect numbers.

On March 26, 1936, the *New York Herald Tribune* published an article that claimed that Dr. Samuel I. Krieger discovered a perfect number with more than 19 digits. Although mathematicians previously dispelled this notion, it was not until 1952 that, with the aid of a computer, the number $2^{257} - 1$ was shown to be a composite number, and, therefore, not capable of generating a perfect number. Mathematics journals chided the newspaper for the eagerness to report a story before verifying its accuracy.

### Here's the Numbers Problem Euclid Tried to Organize

Chicago Scientist Proves That 155 Digits Are Perfect—Result of 5 Years' Work

CHICAGO, March 26, 1936—Dr. Samuel I. Krieger laid down his pencil and paper today and claimed he had solved a problem that had baffled mathematicians since Euclid's day—the finding of a perfect number of more than nineteen digits.

A perfect number is one that is equal to the sum of its divisors, he explained. For example, 28 is the sum of 1, 2, 4, 7 and 14, all of which may be divided into it.

Dr. Krieger's perfect number contains 155 digits. Here it is:
26,815,615,859,885,194,199,148,049,996,411,692,254,958,731,641,184,78
6,755,447,122,887,443,528,060,146,978,161,514,511,280,138,383,284,395,
055,028,465,118,831,722,842,125,059,853,682,308,859,384,882,528,256.

Its formula is two to the 513th power minus two to the 256th power.

The doctor said it took him seventeen hours to work it out and five years to prove it correct.

Mathematicians are always fascinated with perfect numbers, and consequently are always in search for further members of this set of perfect numbers. As of the publication date, there are more than 50 known perfect numbers today, We show the first 48 perfect numbers in Figure 71.

| Rank | $n$ | Perfect number | Digits | Year discovered |
|------|-----|----------------|--------|-----------------|
| 1 | 2 | 6 | 1 | Greeks |
| 2 | 3 | 28 | 2 | Greeks |
| 3 | 5 | 496 | 3 | Greeks |
| 4 | 7 | 8128 | 4 | Greeks |
| 5 | 13 | 33550336 | 8 | 1456 |
| 6 | 17 | 8589869056 | 10 | 1588 |
| 7 | 19 | 137438691328 | 12 | 1588 |
| 8 | 31 | 2305843008139952128 | 19 | 1772 |
| 9 | 61 | 2658455991569831744654692615953842176 | 37 | 1883 |
| 10 | 89 | 191561942608236107294793378084303638130997321548169216 | 54 | 1911 |
| 11 | 107 | 131640364...783728128 | 65 | 1914 |
| 12 | 127 | 144740111...199152128 | 77 | 1876 |
| 13 | 521 | 235627234...555646976 | 314 | 1952 |
| 14 | 607 | 141053783...537328128 | 366 | 1952 |
| 15 | 1279 | 541625262...984291328 | 770 | 1952 |
| 16 | 2203 | 108925835...453782528 | 1327 | 1952 |
| 17 | 2281 | 994970543...139915776 | 1373 | 1952 |
| 18 | 3217 | 335708321...628525056 | 1937 | 1957 |
| 19 | 4253 | 182017490...133377536 | 2561 | 1961 |
| 20 | 4423 | 407672717...912534528 | 2663 | 1961 |
| 21 | 9689 | 114347317...429577216 | 5834 | 1963 |
| 22 | 9941 | 598885496...073496576 | 5985 | 1963 |
| 23 | 11213 | 395961321...691086336 | 6751 | 1963 |
| 24 | 19937 | 931144559...271942656 | 12003 | 1971 |
| 25 | 21701 | 100656497...141605376 | 13066 | 1978 |
| 26 | 23209 | 811537765...941666816 | 13973 | 1979 |
| 27 | 44497 | 365093519...031827456 | 26790 | 1979 |
| 28 | 86243 | 144145836...360406528 | 51924 | 1982 |
| 29 | 110503 | 136204582...603862528 | 66530 | 1988 |
| 30 | 132049 | 131451295...774550016 | 79502 | 1983 |
| 31 | 216091 | 278327459...840880128 | 130100 | 1985 |
| 32 | 756839 | 151616570...565731328 | 455663 | 1992 |
| 33 | 859433 | 838488226...416167936 | 517430 | 1994 |
| 34 | 1257787 | 849732889...118704128 | 757263 | 1996 |
| 35 | 1398269 | 331882354...723375616 | 841842 | 1996 |
| 36 | 2976221 | 194276425...174462976 | 1791864 | 1997 |
| 37 | 3021377 | 811686848...022457856 | 1819050 | 1998 |
| 38 | 6972593 | 955176030...123572736 | 4197919 | 1999 |
| 39 | 13466917 | 427764159...863021056 | 8107892 | 2001 |
| 40 | 20996011 | 793508909...206896128 | 12640858 | 2003 |

**Figure 71** *(continued)*

| 41 | 24036583 | 448233026...572950528 | 14471465 | 2004 |
| 42 | 25964951 | 746209841...791088128 | 15632458 | 2005 |
| 43 | 30402457 | 497437765...164704256 | 18304103 | 2005 |
| 44 | 32582657 | 775946855...577120256 | 19616714 | 2006 |
| 45 | 37156667 | 204534225...074480128 | 22370543 | 2008 |
| 46 | 42643801 | 144285057...377253376 | 25674127 | 2009 |
| 47 | 43112609 | 500767156...145378816 | 25956377 | 2008 |
| 48 | 57885161 | 169296395...270130176 | 34850340 | 2013 |

**Figure 71**

There are lots of curious characteristics of perfect numbers. For example, we notice that they all end in either 28 or 6, and that preceded by an odd digit. Mathematicians have been in search for odd perfect numbers, and as we can see our list consists only of even perfect numbers. To date, they say, with confidence, that there are no odd perfect numbers less than $10^{1500}$.

There are many unusual characteristics of perfect numbers beyond those that define them—namely, that the sum of their divisors equals the number itself. Here are some curious characteristics of perfect numbers of the form $2^{n-1}(2^n - 1)$.

First, they are the sum of the first consecutive natural numbers as shown in the following equation:

$6 = 1 + 2 + 3$

$28 = 1 + 2 + 3 + 4 + 5 + 6 + 7$

$496 = 1 + 2 + 3 + 4 + 5 + 6 + 7 + 8 + 9 + 10 + 11 + 12 + 13 + 14 + \ldots$
$\qquad + 29 + 30 + 31$

$8,128 = 1 + 2 + 3 + 4 + 5 + 6 + 7 + 8 + 9 + 10 + 11 + 12 + 13 + 14 + \ldots$
$\qquad + 125 + 126 + 127$

$33,550,336 = 1 + 2 + 3 + 4 + 5 + 6 + 7 + 8 + 9 + 10 + 11 + 12 + 13$
$\qquad + \ldots + 8,189 + 8,190 + 8,191$

and so on.

If this isn't enough, we also notice with the exception of the first perfect number, 6, they are equal to the sum of consecutive odd cubes as shown in the equation that follows:

$28 = 1^3 + 3^3$

$496 = 1^3 + 3^3 + 5^3 + 7^3$

$8,128 = 1^3 + 3^3 + 5^3 + 7^3 + 9^3 + 11^3 + 13^3 + 15^3$

$33,550,336 = 1^3 + 3^3 + 5^3 + \ldots + 123^3 + 125^3 + 127^3$

Now, you may wonder, how can it be that suddenly perfect numbers are equal to the sum of the cubes of a sequence of odd numbers? This can be easily justified through some elementary algebra. We recall that the perfect number must take the form of $2^{n-1}(2^n - 1)$, where $2^n - 1$ is a prime number. We will take a moment here to show how we can justify that each perfect number is the sum of the first $2^k$ odd numbers, where $k = \dfrac{1}{2}(n - 1)$, except for $n = 2$. We should recall that the following are true:

$$S_1 = 1 + 2 + 3 + 4 + \quad + q = \frac{q(q+1)}{2}$$

$$S_2 = 1^2 + 2^2 + 3^2 + 4^2 + \quad + q^2 = \frac{q(q+1)(2q+1)}{6}$$

$$S_3 = 1^3 + 2^3 + 3^3 + 4^3 + \quad + q^3 = \frac{q^2(q+1)^2}{4} = S_1^2$$

**Equation 20**

Let us now look at the sum of the cubes of the odd numbers. We can write that as follows:

$$S = 1^3 + \ 3^3 + \ 5^3 + 7^3 + \ldots + (2q - 1) = \sum_{i=1}^{q}(2q - 1)^3.$$

With some algebraic manipulation[2] we can show that this is equal to the following:

$$S = \sum_{i=1}^{q}(2q - 1)^3 = \sum_{i=1}^{q}(8q^3 - 12q^2 + 6q - 1)$$

$$= 8 \cdot \frac{q^2(q+1)^2}{4} - 12 \cdot \frac{q(q+1)(2q+1)}{6} + 6 \cdot \frac{q(q+1)}{2} - q = q^2(2q^2 - 1)$$

Notice how we inserted the first three formulas, which we generated preceding information, into this last equation. If we now let $q = 2^k$, then $S = 2^{2k}(2^{2k+1} - 1)$, then we notice that this would generate the perfect numbers, $2^{n-1}(2^n - 1)$, when $n = 2k + 1$, which are the odd numbers. This, then, shows how we go from the general formula for the perfect numbers to the sum of the cubes of the odd numbers. Although this may be a little bit of some tough elementary algebra, we provided it here to show how we can justify some of these mathematical curiosities.

All perfect numbers must have an even number of divisors. If we take the reciprocals of the divisors of any perfect number (now, including the number itself), their sum will always be equal to 2. This can be seen from the first few perfect numbers as follows:

$$\frac{1}{1}+\frac{1}{2}+\frac{1}{3}+\frac{1}{6}=2$$

$$\frac{1}{1}+\frac{1}{2}+\frac{1}{4}+\frac{1}{7}+\frac{1}{14}+\frac{1}{28}=2$$

$$\frac{1}{1}+\frac{1}{2}+\frac{1}{4}+\frac{1}{8}+\frac{1}{16}+\frac{1}{31}+\frac{1}{62}+\frac{1}{124}+\frac{1}{248}+\frac{1}{496}=2$$

**Equation 21**

Although much of the properties of the perfect numbers discussed earlier is once again covered here, we are offering a deeper understanding of the concepts involved to provide greater enrichment.

## KAPREKAR NUMBERS

We initially encountered Kaprekar numbers with a number 297, where we found that $297^2 = 88,209$, and $88 + 209 = 297$. Such a number is called a *Kaprekar number*, named after the Indian mathematician Dattaraya Ramchandra Kaprekar (1905–1986) who discovered such numbers. Some other Kaprekar numbers are shown in the table in Figure 72.

Other Kaprekar numbers are: 38,962; 77,778; 82656; 95,121; 99,999; 142,857; ...; 538,461; 857,143; .... We also have such numbers as Kaprekar triples, which behave as follows: $45^3 = 91,125 = 9 + 11 + 25 = 45$. Some other Kaprekar triples are: 1, 8, 10, 297, 2,322.

Earlier, we showed that 297 is a Kaprekar number, and now we show how it is also a Kaprekar triple: $297^3 = 26,198,073$, and $26 + 198 + 073 = 297$.

It should also be noted that the smallest 10-digit Kaprekar number is 1,111,111,111, which is a repunit number, which is to be considered in the next unit. By the way, the square of this number is 1,234,567,900,987,654,321, where the sum of its parts is $123,456,790 + 0,987,654,321 = 1,111,111,111$, thus making it a Kaprekar number. Furthermore, the number 22,222,222,222,222 is also a Kaprekar number because $22,222,222,222,222^2 = 493,827,160,493,817,283,950,617,284$. Then $4,938,271,604,938 + 17,283,950,617,284 = 22,222,222,222,222$.

| Kaprekar numbers | | Square of the numbers | | | Decomposition of the numbers |
|---|---|---|---|---|---|
| 1 | | $1^2$ | = | 1 | $1 = 1$ |
| 9 | | $9^2$ | = | 81 | $8 + 1 = 9$ |
| 45 | | $45^2$ | = | 2,025 | $20 + 25 = 45$ |
| 55 | | $55^2$ | = | 3,025 | $30 + 25 = 55$ |
| 99 | | $99^2$ | = | 9,801 | $98 + 01 = 99$ |
| 297 | | $297^2$ | = | 88,209 | $88 + 209 = 297$ |
| 703 | | $703^2$ | = | 494,209 | $494 + 209 = 703$ |
| 999 | | $999^2$ | = | 998,001 | $998 + 001 = 999$ |
| 2,223 | | $2,223^2$ | = | 4,941,729 | $494 + 1,729 = 2,223$ |
| 2,728 | | $2728^2$ | = | 7,441,984 | $744 + 1,984 = 2,728$ |
| 4,879 | | $4,879^2$ | = | 23,804,641 | $238 + 04,641 = 4,879$ |
| 4,950 | | $4,950^2$ | = | 24,502,500 | $2,450 + 2,500 = 4,950$ |
| 5,050 | | $5,050^2$ | = | 25,502,500 | $2,550 + 2,500 = 5,050$ |
| 5,292 | | $5,292^2$ | = | 28,005,264 | $28 + 005,264 = 5,292$ |
| 7,272 | | $7,272^2$ | = | 52,881,984 | $5,288 + 1,984 = 7,272$ |
| 7,777 | | $7,777^2$ | = | 60,481,729 | $6,048 + 1,729 = 7,777$ |
| 9,999 | | $999^2$ | = | 99,980,001 | $9,998 + 0,001 = 9,999$ |
| 17,344 | | $17,344^2$ | = | 300,814,336 | $3,008 + 14,336 = 17,344$ |
| 22,222 | | $22,222^2$ | = | 493,817,284 | $4,938 + 17,284 = 22,222$ |

**Figure 72**

## THE REPUNIT NUMBERS (WHERE ALL THE DIGITS ARE ONES)

Having earlier seen some of the unusual features of the number 11, we will consider large numbers consisting of only 1s—called *repunits*.[3] (Sometimes also referred to as unit-digit numbers.) The next larger number after 11 that consists of all 1s is the number 111, and it, too, has some curious properties. For example, the differences of two squares, can be so selected equal to a number consisting of all 1s. The progression of such differences of squares leading to numbers consisting of repunit numbers follows:

$$1^2 - 0^2 = 1$$

$$6^2 - 5^2 = 11$$

$$20^2 - 17^2 = 111$$

$$56^2 - 45^2 = 1,111$$

$$156^2 - 115^2 = 11,111$$

$$556^2 - 445^2 = 111,111$$

$$344^2 - 85^2 = 111,111$$

$$356^2 - 125^2 = 111,111$$

**Equation 22**

Within this list of numbers that will result in 1s, we may see another pattern emerging. If we look at the second, fourth, and sixth entries, we will notice an additional pattern between the generating numbers from one to another. Each time an additional 5 and 4 is tagged onto the front of the numbers, respectively. If we continue this pattern notice what a spectacular pattern evolves.

$$6^2 - 5^2 = 11$$
$$56^2 - 45^2 = 1,111$$
$$556^2 - 445^2 = 111,111$$
$$5556^2 - 4445^2 = 11,111,111$$
$$55556^2 - 44445^2 = 1,111,111,111$$
$$555556^2 - 444445^2 = 111,111,111,111$$
$$5555556^2 - 4444445^2 = 11,111,111,111,111$$
$$55555556^2 - 44444445^2 = 1,111,111,111,111,111$$
$$555555556^2 - 444444445^2 = 111,111,111,111,111,111$$

...

$$555555555555555556^2 - 444444444444444445^2 =$$

$$1,111,111,111,111,111,111,111,111,111,111,111,111$$

**Equation 23**

Of this list, the only prime number is 11. In fact, the next two prime numbers of all 1s are:

1,111,111,111,111,111,111 and 11,111,111,111,111,111,111,111. It is quite obvious that these last two numbers will be prime in any arrangement of the digits—because all the digits are the same!

However, we should be aware that there are, in fact, prime numbers, where all arrangements of their digits result in another prime number. The first few of these are 11, 13, 17, 37, 79, 113, 199, and 337. You might like to find the next few such primes that allow new prime numbers with all arrangements of their digits.

Then there are more curious numbers; those, that are generated by the difference of squares and are multiples of numbers consisting of digits that are all 1s.

$$7^2 - 4^2 = 33 = 3 \times 11$$
$$67^2 - 34^2 = 3,333 = 3 \times 1,111$$
$$667^2 - 334^2 = 333,333 = 3 \times 111,111$$
$$6667^2 - 3334^2 = 33,333,333 = 3 \times 11,111,111$$
$$66667^2 - 33334^2 = 3,333,333,333 = 3 \times 1,111,111,111$$

**Equation 24**

Here is another such pattern of numbers, which should be admired.

$$8^2 - 3^2 = 55 = 5 \times 11$$
$$78^2 - 23^2 = 5555 = 5 \times 1,111$$
$$778^2 - 223^2 = 555,555 = 5 \times 111,111$$
$$7778^2 - 2223^2 = 55,555,555 = 5 \times 11,111,111$$
$$77778^2 - 22223^2 = 5,555,555,555 = 5 \times 1,111,111,111$$

**Equation 25**

Our further investigation of repunit numbers brings us to an interesting pattern, one where we divide 111,111,111 by 9, which gives us the number 12,345,679. Notice we have the digits in numerical order, but we are missing the number 8. Yet, when we consider the following pattern the 8 is once again included in generating numbers consisting of only 1s.

| | | | | | | |
|---:|:---:|:---:|:---:|:---:|:---:|:---|
| 0 | × | 9 | + | 1 | = | 1 |
| 1 | × | 9 | + | 2 | = | 11 |
| 12 | × | 9 | + | 3 | = | 111 |
| 123 | × | 9 | + | 4 | = | 1,111 |
| 1,234 | × | 9 | + | 5 | = | 11,111 |
| 12,345 | × | 9 | + | 6 | = | 111,111 |
| 123,456 | × | 9 | + | 7 | = | 1,111,111 |
| 1,234,567 | × | 9 | + | 8 | = | 11,111,111 |
| 12,345,678 | × | 9 | + | 9 | = | 111,111,111 |
| 123,456,789 | × | 9 | + | 10 | = | 1,111,111,111 |

**Equation 26**

As you can see, repunit numbers seem to generate some rather interesting relationships and patterns. Another curious pattern evolves, when we take the square of successive unit-digit numbers as shown in Figure 73.

| Number of 1's | $n$ | $n^2$ |
|---|---|---|
| 1 | 1 | 1 |
| 2 | 11 | 121 |
| 3 | 111 | 12321 |
| 4 | 1111 | 1234321 |
| 5 | 11111 | 123454321 |
| 6 | 111111 | 12345654321 |
| 7 | 1111111 | 1234567654321 |
| 8 | 11111111 | 123456787654321 |
| 9 | 111111111 | 12345678987654321 |
| 10 | 1111111111 | 1234567900987654321 |

**Figure 73**

To get a better view of these repunit numbers, $r_n$, we will factor them into their prime factors as shown in Figure 74.

We notice that $r_2$ and $r_{19}$ are prime numbers. The question then arises are there other such repunits, which are primes? The answer is yes. However, mathematicians have struggled with this question for years. For example, the German mathematician Gustav Jacob Jacobi (1804–1851) pursued the question as to whether the repunit $r_{11}$ is a prime number. Today, a computer algebra system can answer this question in less than a second. Factoring repunit numbers is often very difficult, however with the aid of a computer we find that the repunit $r_{71}$ is factorable as follows, and, therefore, not a prime number.

$r_{71}$ = 11,111,111,111,111,111,111,111,111,111,111,111,111,111,111,111, 111,111,111,111,111,111,111,111 = 241,573,142,393,627,673,576,957,439,049 × 45,994,811,347,886,846,310,221,728,895,223,034,301,839.

By 1930, it was known that $r_2$ and $r_{19}$ (Oscar Hoppe, 1916) as well as $r_{23}$ (Lehmer and Kraitchik, 1929) are prime numbers.[4] In 1970, a mathematics student, E. Seah, was able to demonstrate that the repunit $r_{317}$ is also a prime number. The search for a prime number among the repunits continues. For example, in 1985 the repunit $r_{1031}$ was discovered to be a prime by H. C. Williams and H. Dubner. Further repunits that have been since identified as primes are $r_{49081}$ (H. Dubner, 1999), $r_{86453}$ (L. Baxter, 2000), $r_{109297}$ (P. Bourdelais, H. Dubner, 2007), and $r_{270343}$ (M. Voznyy, A. Budnyy, 2007).[5] It is believed today that there are an infinite number of repunits that are prime numbers.

| | | | | |
|---|---|---|---|---|
| $r_1$ | = | 1 | = | 1 |
| $r_2$ | = | 11 | = | 11 |
| $r_3$ | = | 111 | = | 3×37 |
| $r_4$ | = | 1,111 | = | 11×101 |
| $r_5$ | = | 11,111 | = | 41×271 |
| $r_6$ | = | 111,111 | = | 3×7×11×13×37 |
| $r_7$ | = | 1,111,111 | = | 239×4,649 |
| $r_8$ | = | 11,111,111 | = | 11×73×101×137 |
| $r_9$ | = | 111,111,111 | = | $3^2$×37×333,667 |
| $r_{10}$ | = | 1,111,111,111 | = | 11×41×271×9,091 |
| $r_{11}$ | = | 11,111,111,111 | = | 21,649×513,239 |
| $r_{12}$ | = | 111,111,111,111 | = | 3×7×11×13×37×101×9,901 |
| $r_{13}$ | = | 1,111,111,111,111 | = | 53×79×265,371,653 |
| $r_{14}$ | = | 11,111,111,111,111 | = | 11×239×4,649×909,091 |
| $r_{15}$ | = | 111,111,111,111,111 | = | 3×31×37×41×271×2,906,161 |
| $r_{16}$ | = | 1,111,111,111,111,111 | = | 11×17×73×101×137×5,882,353 |
| $r_{17}$ | = | 11,111,111,111,111,111 | = | 2,071,723×5,363,222,357 |
| $r_{18}$ | = | 111,111,111,111,111,111 | = | $3^2$×7×11×13×19×37×52,579×333,667 |
| $r_{19}$ | = | **1,111,111,111,111,111,111** | = | **1,111,111,111,111,111,111** |
| $r_{20}$ | = | 11,111,111,111,111,111,111 | = | 11×41×101×271×3,541×9,091×27,961 |

**Figure 74**

## SOME NUMBER PECULIARITIES

The representation of all nine digits often fascinates the observer. Although we have encountered some of these situations earlier in the book, let's reconsider a number of such examples now. One such unexpected result happens when we subtract the symmetric numbers consisting of the digits in consecutive reverse order and in numerical order:

$987,654,321 - 123,456,789$ to get 864,197,532. This symmetric subtraction used each of the nine digits exactly once in each of the numbers being subtracted, and surprisingly, resulted in a difference that also used each of the nine digits exactly once.

Here are a few more such strange calculations—this time using multiplication—where on either side of the equal sign all nine digits are represented exactly once: $291,548,736 = 8 \times 92 \times 531 \times 746$, and also for $124,367,958 = 627 \times 198,354 = 9 \times 26 \times 531,487$.

Another example of a calculation, where all the digits are used exactly once (not counting the exponent), is $567^2 = 321,489$. This also works for the following: $854^2 = 729,316$. These are, apparently, the only two squares that result in a number, which allow all the digits to be represented once.

A somewhat convoluted calculation that results in a surprise ending begins with the following: $6,667^2 = 44,448,889$. When this result, 44,448,889, is multiplied by 3 to get 133,346,667, we notice that the last four digits are the same as the four digits of the number we began with, namely, 6,667.

Number oddities are boundless. Some of these seem a bit far-fetched, but nonetheless can be appealing to us from a recreational point of view. For example, consider taking any three-digit number that is multiplied by a five-digit number, all of whose digits are the same. When you add its last five digits to the remaining digits, a number will result, where all digits are the same. Here are a few such examples:

$237 \times 33,333 = 7,899,921$, then $78 + 99,921 = 99,999$.
$357 \times 77,777 = 27,766,389$, then $277 + 66,389 = 66,666$.
$789 \times 44,444 = 35,066,316$, then $350 + 66,316 = 66,666$.
$159 \times 88,888 = 14,133,192$, then $141 + 33,192 = 33,333$.

These amazing number peculiarities, although entertaining, allow us to exhibit the beauty of mathematics!

## ARMSTRONG NUMBERS

As we continue to expose some of the most celebrated numbers in mathematics, we come to those that are earlier referred to as Armstrong numbers (named after the mathematician Michael F. Armstrong) or sometimes referred to as narcissistic numbers. The *Armstrong numbers* have the property that each number is equal to the sum of its digits, when each is taken to the power equal to the number of digits in the original number. For example, we have earlier encountered the three-digit Armstrong number $153 = 1^3 + 5^3 + 3^3 = 1 + 125 + 27$.

A nine-digit number can also be an Armstrong number because $472,335,975 = 4^9 + 7^9 + 2^9 + 3^9 + 3^9 + 5^9 + 9^9 + 7^9 + 5^9$.

All Armstrong numbers are shown in the Figure 75, where we notice that there are no Armstrong numbers for $k = 2, 12, 13, 15, 18, 22, 26, 28, 30$, and $36$ (and $k > 39$). In fact, there are only 89 Armstrong numbers in the decimal system. The largest Armstrong number is 39 digits long, and it is equal to the sum of its digits, each of which is taken to the 39th power:

$$1^{39} + 1^{39} + 5^{39} + 1^{39} + 3^{39} + 2^{39} + 2^{39} + 1^{39} + 9^{39} + 0^{39} + 1^{39} + 8^{39} + 7^{39}$$
$$+ 6^{39} + 3^{39} + 9^{39} + 9^{39} + 2^{39} + 5^{39} + 6^{39} + 5^{39} + 0^{39} + 9^{39} + 5^{39} + 5^{39} + 9^{39}$$
$$+ 7^{39} + 9^{39} + 7^{39} + 3^{39} + 9^{39} + 7^{39} + 1^{39} + 5^{39} + 2^{39} + 2^{39} + 4^{39} + 0^{39} + 1^{39}$$
$$= 115{,}132{,}219{,}018{,}763{,}992{,}565{,}095{,}597{,}973{,}971{,}522{,}401.$$

| no. | digits | Armstrong Number | no. | digits | Armstrong Number |
|---|---|---|---|---|---|
| 0 | 1 | 0 | 45 | 17 | 35641594208964132 |
| 1 | 1 | 1 | 46 | 17 | 35875699062250035 |
| 2 | 1 | 2 | 47 | 19 | 1517841543307505039 |
| 3 | 1 | 3 | 48 | 19 | 3289582984443187032 |
| 4 | 1 | 4 | 49 | 19 | 4498128791164624869 |
| 5 | 1 | 5 | 50 | 19 | 4929273885928088826 |
| 6 | 1 | 6 | 51 | 20 | 63105425988599693916 |
| 7 | 1 | 7 | 52 | 21 | 128468643043731391252 |
| 8 | 1 | 8 | 53 | 21 | 449177399146038697307 |
| 9 | 1 | 9 | 54 | 23 | 21887696841122916288858 |
| 10 | 3 | 153 | 55 | 23 | 27879694893054074471405 |
| 11 | 3 | 370 | 56 | 23 | 27907865009977052567814 |
| 12 | 3 | 371 | 57 | 23 | 28361281321319229463398 |
| 13 | 3 | 407 | 58 | 23 | 35452590104031691935943 |
| 14 | 4 | 1634 | 59 | 24 | 174088005938065293023722 |
| 15 | 4 | 8208 | 60 | 24 | 188451485447897896036875 |
| 16 | 4 | 9474 | 61 | 24 | 239313664430041569350093 |
| 17 | 5 | 54748 | 62 | 25 | 1550475334214501539088894 |
| 18 | 5 | 92727 | 63 | 25 | 1553242162893771850669378 |
| 19 | 5 | 93084 | 64 | 25 | 3706907995955475988644380 |
| 20 | 6 | 548834 | 65 | 25 | 3706907995955475988644381 |
| 21 | 7 | 1741725 | 66 | 25 | 4422095118095899619457938 |
| 22 | 7 | 4210818 | 67 | 27 | 121204998563613372405438066 |
| 23 | 7 | 9800817 | 68 | 27 | 121270696006801314328439376 |
| 24 | 7 | 9926315 | 69 | 27 | 128851796696487777842012787 |
| 25 | 8 | 24678050 | 70 | 27 | 174650464499531377631639254 |
| 26 | 8 | 24678051 | 71 | 27 | 177265453171792792366489765 |
| 27 | 8 | 88593477 | 72 | 29 | 14607640612971980372614873089 |
| 28 | 9 | 146511208 | 73 | 29 | 19008174136254279950127347 40 |
| 29 | 9 | 472335975 | 74 | 29 | 19008174136254279950127347 41 |
| 30 | 9 | 534494836 | 75 | 29 | 23866716435523975980390369295 |
| 31 | 9 | 912985153 | 76 | 31 | 1145037275765491025924292050346 |
| 32 | 10 | 4679307774 | 77 | 31 | 1927890457142960697580636236639 |
| 33 | 11 | 32164049650 | 78 | 31 | 2309092682616190307509695338915 |
| 34 | 11 | 32164049651 | 79 | 32 | 17333509997782249308725103962772 |
| 35 | 11 | 40028394225 | 80 | 33 | 186709961001538790100634132976990 |
| 36 | 11 | 42678290603 | 81 | 33 | 186709961001538790100634132976991 |
| 37 | 11 | 44708635679 | 82 | 34 | 1122763285329372541592822900204593 |
| 38 | 11 | 49388550606 | 83 | 35 | 12639369517103790328947807201478392 |
| 39 | 11 | 82693916578 | 84 | 35 | 12679937780272278566303885594196922 |
| 40 | 11 | 94204591914 | 85 | 37 | 1219167219625434121569735803609966019 |
| 41 | 14 | 28116440335967 | 86 | 38 | 12815792078366059955099770545296129367 |
| 42 | 16 | 4338281769391370 | 87 | 39 | 115132219018763992565095597973971522400 |
| 43 | 16 | 4338281769391371 | 88 | 39 | 115132219018763992565095597973971522401 |
| 44 | 17 | 21897142587612075 | | | |

**Figure 75**

The following is a list of the *consecutive* Armstrong numbers.

$k = 3$:          370;    371;

$k = 8$:          24,678,050;    24,678,051;

$k = 11$:         32,164,049,650;    32,164,049,651;

$k = 16$:         4,338,281,769,391,370;    4,338,281,769,391,371;

$k = 25$:         3,706,907,995,955,475,988,644,380;    3,706,907,995,955,475,988,644,381;

$k = 29$:         19,008,174,136,254,279,995,012,734,740;

                  19,008,174,136,254,279,995,012,734,741;

$k = 33$:         186,709,961,001,538,790,100,634,132,976,990;

                  186,709,961,001,538,790,100,634,132,976,991;

$k = 39$:         115,132,219,018,763,992,565,095,597,973,971,522,400;

                  115,132,219,018,763,992,565,095,597,973,971,522,401

**Equation 27**

Incidentally, our first Armstrong number, 153, has some other amazing properties, which we mentioned earlier: it is a triangular number, where we have the sum of the first 17 numbers equal to 153. $1 + 2 + 3 + 4 + 5 + 6 + 7 + 8 + 9 + 10 + 11 + 12 + 13 + 14 + 15 + 16 + 17 = 153$.

The number 153 is not only equal to the sum of the cubes of its digits, but it is also a number that can be expressed as the sum of consecutive factorials: $1! + 2! + 3! + 4! + 5! = 153$.

## HAPPY AND UNHAPPY NUMBERS

It is not clear from where the term *happy number* evolved. Most people feel that the British mathematician Reg Allenby's daughter came home from school after recognizing this unusual relationship and so call them happy numbers. We say that a number is considered a *happy number* if we take the sum of the squares of the digits to get a second number, and then take the sum of the squares of the digits of that second number to get a third number, and then take the sum of the squares of the digits of the third number, and then continue until you get to the number 1, which then allows us to refer to this original number as a happy number. The numbers, where this scheme is applied and do *not* eventually end with the number 1 are called *unhappy numbers*. These latter numbers will eventually end up in a loop—where they will continue through a cycle.

Let's consider one of the happy numbers, say 13, and follow this process of taking the sum of the squares of the digits continuously.

$$1^2 + 3^2 = 10$$
$$1^2 + 0^2 = 1$$

Let's consider a slightly longer path to the number 1, by using 19.

$$1^2 + 9^2 = 82$$
$$8^2 + 2^2 = 68$$
$$6^2 + 8^2 = 100$$
$$1^2 + 0^2 + 0^2 = 1$$

Following is a list of the happy numbers from 1 to 1,000. You might want to try a few of these to test for their "happiness."

1, 7, 10, 13, 19, 23, 28, 31, 32, 44, 49, 68, 70, 79, 82, 86, 91, 94, 97, 100, 103, 109, 129, 130, 133, 139, 167, 176, 188, 190, 192, 193, 203, 208, 219, 226, 230, 236, 239, 262, 263, 280, 291, 293, 301, 302, 310, 313, 319, 320, 326, 329, 331, 338, 356, 362, 365, 367, 368, 376, 379, 383, 386, 391, 392, 397, 404, 409, 440, 446, 464, 469, 478, 487, 490, 496, 536, 556, 563, 565, 566, 608, 617, 622, 623, 632, 635, 637, 638, 644, 649, 653, 655, 656, 665, 671, 673, 680, 683, 694, 700, 709, 716, 736, 739, 748, 761, 763, 784, 790, 793, 802, 806, 818, 820, 833, 836, 847, 860, 863, 874, 881, 888, 899, 901, 904, 907, 910, 912, 913, 921, 923, 931, 932, 937, 940, 946, 964, 970, 973, 989, 998, 1,000.

The first few pairs of consecutive happy numbers $(n; n + 1)$ are: (31, 32), (129, 130), (192, 193), (262, 263), (301, 302), (319, 320), (367, 368), (391, 392), …

As you will see when you try to discover happy numbers, many will fall into similar paths, such as the numbers 19 and 91. Following are the unique combinations, that is, none of these numbers will take the same path as they get to the number 1. Naturally, the preceding list includes variations of these numbers: 1, 7, 13, 19, 23, 28, 44, 49, 68, 79, 129, 133, 139, 167, 188, 226, 236, 239, 338, 356, 367, 368, 379, 446, 469, 478, 556, 566, 888, 899.

It appears that happy numbers are without zeros and with digits in increasing order.

An example of an unhappy number is 25. Notice how this procedure will lead us into a loop and never to reach the number 1.

$$2^2 + 5^2 = 29$$
$$2^2 + 9^2 = 85$$
$$8^2 + 5^2 = 89$$
$$8^2 + 9^2 = 145$$
$$1^2 + 4^2 + 5^2 = 42$$
$$4^2 + 2^2 = 20$$
$$2^2 + 0^2 = 4$$
$$4^2 = 16$$
$$1^2 + 6^2 = 37$$
$$3^2 + 7^2 = 58$$
$$5^2 + 8^2 = 89$$

**Equation 28**

This then begins another loop—repeating everything again with 89, which was reached earlier in this process.

Among the happy numbers there are those that are prime numbers as well—known as *happy prime numbers*. Those numbers less than 500 are the following: 7, 13, 19, 23, 31, 79, 97, 103, 109, 139, 167, 193, 239, 263, 293, 313, 331, 367, 379, 383, 397, 409, 487.

To date, mathematicians have made some claims as to some special happy numbers. For example, the smallest happy number that exhibits all the digits 0 to 9 is 10,234,456,789. While the smallest happy number that has no zeros and yet has all the other digits is 1,234,456,789. Then there is the smallest happy number that has no zeros and is also palindromic: 13,456,789,298,765,431. When we accept the inclusion of zeros, then the smallest happy number with all the digits is 1,034,567,892,987,654,301. Thus far, the largest happy number where no digit is repeated is 986,543,210.

We recall the 2nd and 3rd repunit primes, $r_{19} = 1,111,111,111,111,111,111$ and $r_{23} = 11,111,111,111,111,111,111,111$, which also happen to be happy primes.

An ambitious reader may find other special characteristics among the happy numbers.

## SOME PRIME DENOMINATORS

Earlier we have considered prime numbers, those that have no factors other than themselves and 1, but now we will consider the fractions whose denominators are prime numbers (excluding 2 and 5), and where the decimal expansion yields an even number of repetitions. These will give us further material to discover patterns to amaze us. Some of these are as follows:

$$\frac{1}{7} = .142857142857142857 \quad = .\overline{142857}$$

$$\frac{1}{11} = .090909 \quad = .\overline{09}$$

$$\frac{1}{13} = .\overline{076923}$$

$$\frac{1}{17} = .\overline{0588235294117647}$$

$$\frac{1}{19} = .\overline{052631578947368421}$$

$$\frac{1}{23} = 0.\overline{0434782608695652173913}$$

**Equation 29**

We will now treat each one of these repeating periods as a number and split this even-digit sequence of digits into two parts, and then add them. An amazing result[6] appears, as shown in Figure 76.

| | |
|---|---|
| $\frac{1}{7} = 0.\overline{142857}$ | 142<br>857<br>999 |
| | |
| $\frac{1}{11} = 0.\overline{09}$ | 0<br>9<br>9 |
| | |
| $\frac{1}{13} = 0.\overline{076923}$ | 076<br>923<br>999 |
| | |
| $\frac{1}{17} = 0.\overline{0588235294117647}$ | 05882352<br>94117647<br>99999999 |
| | |
| $\frac{1}{19} = 0.\overline{052631578947368421}$ | 052631578<br>947368421<br>999999999 |
| | |
| $\frac{1}{23} = 0.\overline{0434782608695652173913}$ | 04347826086<br>95652173913<br>99999999999 |

**Figure 76**

By the way, speaking of prime numbers, the numbers 139 and 149 are the first two prime numbers that differ by 10.

## SOME MORE CURIOUS NUMBERS

We will now embark on a most unusual numerical phenomenon. This time we will begin by considering unit fractions whose denominators are a multiple of 9, and not a multiple of 2 or of 5. Some examples of this would be the following fractions: $\frac{1}{27}, \frac{1}{63}, \frac{1}{81}, \frac{1}{99}, \frac{1}{117}, \frac{1}{153}$, and $\frac{1}{171}$. When we convert each of these fractions to decimals form—by dividing the denominator into the numerator—we get the following decimals:

$$\frac{1}{9 \times 3} = \frac{1}{27} = 0.037037037037037037037037037037037 = 0.\overline{037}$$

$$\frac{1}{9 \times 7} = \frac{1}{63} = 0.0158730158730158730158730158730158730158 = 0.\overline{015873}$$

$$\frac{1}{9 \times 9} = \frac{1}{81} = 0.012345679012345679012345679 = 0.\overline{012345679}$$

$$\frac{1}{9 \times 11} = \frac{1}{99} = 0.01010101010101010101010101010101 = 0.\overline{01}$$

$$\frac{1}{9 \times 13} = \frac{1}{117} = 0.008547008547008547008547008547 = 0.\overline{008547}$$

$$\frac{1}{9 \times 17} = \frac{1}{153} = 0.00653594771241830065359477124183 = 0.\overline{0065359477124183}$$

$$\frac{1}{9 \times 19} = \frac{1}{171} = 0.005847953216374269005847953216374269 = 0.\overline{005847953216374269}$$

**Equation 30**

Thus far nothing spectacular has really appeared with this decimal conversion of the unit fractions. What follows is truly an amazing curiosity of our number system, as we have seen in our earlier discussions of specific numbers. Follow along as we will recall how taking the repeating digits (without the initial zeros) and multiplying this newly formed number by the multiple of 9 that was used to get these denominators and observe the marvelous pattern that emerges. For example, in the first case we will multiply 37 by multiples of 3 because the denominator 27 was arrived at by multiplying 9 times 3. Therefore, we will multiply 37 by 3, by 6, by 9, and so on, until we reach 27.

$$37 \times 3 = 111$$
$$37 \times 6 = 222$$
$$37 \times 9 = 333$$
$$37 \times 12 = 444$$
$$37 \times 15 = 555$$
$$37 \times 18 = 666$$
$$37 \times 21 = 777$$
$$37 \times 24 = 888$$
$$37 \times 27 = 999$$

**Equation 31**

The next fraction's denominators were attained by taking $9 \times 7$. Now using multiples of 7 and multiplying them by the repeating part, 15,873. Once again, we obtain an easily recognized pattern.

$$15\,873 \times 7 = 111\,111$$
$$15\,873 \times 14 = 222\,222$$
$$15\,873 \times 21 = 333\,333$$
$$15\,873 \times 28 = 444\,444$$
$$15\,873 \times 35 = 555\,555$$
$$15\,873 \times 42 = 666\,666$$
$$15\,873 \times 49 = 777\,777$$
$$15\,873 \times 56 = 888\,888$$
$$15\,873 \times 63 = 999\,999$$

**Equation 32**

Continuing along, we find these results each time delivering these amazing—yet pretty—results for the succeeding fractions listed earlier.

```
12345679 ×  9 = 111111111
12345679 × 18 = 222222222
12345679 × 27 = 333333333
12345679 × 36 = 444444444
12345679 × 45 = 555555555
12345679 × 54 = 666666666
12345679 × 63 = 777777777
12345679 × 72 = 888888888
12345679 × 81 = 999999999
```

```
8 547 ×  13 = 111 111
8 547 ×  26 = 222 222
8 547 ×  39 = 333 333
8 547 ×  52 = 444 444
8 547 ×  65 = 555 555
8 547 ×  78 = 666 666
8 547 ×  91 = 777 777
8 547 × 104 = 888 888
8 547 × 117 = 999 999
```

```
65 359 477 124 183 ×  17 = 1 111 111 111 111 111
65 359 477 124 183 ×  34 = 2 222 222 222 222 222
65 359 477 124 183 ×  51 = 3 333 333 333 333 333
65 359 477 124 183 ×  68 = 4 444 444 444 444 444
65 359 477 124 183 ×  85 = 5 555 555 555 555 555
65 359 477 124 183 × 102 = 6 666 666 666 666 666
65 359 477 124 183 × 119 = 7 777 777 777 777 777
65 359 477 124 183 × 136 = 8 888 888 888 888 888
65 359 477 124 183 × 153 = 9 999 999 999 999 999
```

```
5 847 953 216 374 269 ×  19 = 111 111 111 111 111 111
5 847 953 216 374 269 ×  38 = 222 222 222 222 222 222
5 847 953 216 374 269 ×  57 = 333 333 333 333 333 333
5 847 953 216 374 269 ×  76 = 444 444 444 444 444 444
5 847 953 216 374 269 ×  95 = 555 555 555 555 555 555
5 847 953 216 374 269 × 114 = 666 666 666 666 666 666
5 847 953 216 374 269 × 133 = 777 777 777 777 777 777
5 847 953 216 374 269 × 152 = 888 888 888 888 888 888
5 847 953 216 374 269 × 171 = 999 999 999 999 999 999
```

**Equation 33**

You may want to extend this list of fractions and verify that the pattern continues. Just to show further peculiarities, notice what happens when we reverse the order of one of our previous multiplications (the one evolving from $\frac{1}{81}$). Again, we get an unexpected rather curious pattern!

$$987654321 \times  9 = \mathbf{8}\ 88888888\ \mathbf{9}$$
$$987654321 \times 18 = \mathbf{1\ 7}\ 77777777\ \mathbf{8}$$
$$987654321 \times 27 = \mathbf{2\ 6}\ 66666666\ \mathbf{7}$$
$$987654321 \times 36 = \mathbf{3\ 5}\ 55555555\ \mathbf{6}$$
$$987654321 \times 45 = \mathbf{4\ 4}\ 44444444\ \mathbf{5}$$
$$987654321 \times 54 = \mathbf{5\ 3}\ 33333333\ \mathbf{4}$$
$$987654321 \times 63 = \mathbf{6\ 2}\ 22222222\ \mathbf{3}$$
$$987654321 \times 72 = \mathbf{7\ 1}\ 11111111\ \mathbf{2}$$
$$987654321 \times 81 = \mathbf{8\ 0}\ 00000000\ \mathbf{1}$$

**Equation 34**

There are times when we can arrange numbers—legitimately—creating a rather unexpected pattern as is the case when we find the decimal equivalent of the fraction

$$\frac{1}{729} = \overline{0.0013717421\ 1248285322\ 359364334\ 705075448\ 1618655692}$$
$$\overline{72976680384087971465\ 1989026063\ 1}$$

The 81-period decimal equivalent can be arranged in nine groups of 9 as follows:

```
001 371 742
112 482 853
223 593 964
334 705 075
445 816 186
556 927 297
668 038 408
779 149 519
890 260 631
```

Looking up and down the columns formed by this arrangement will reveal quite a few unusual patterns—such as consecutive numbers. The curiosities in arithmetic seem to be unending!

## CURIOUS PRODUCTS—REVERSALS

It is hard to imagine that there are a certain number pairs that yield the same product even when both numbers are reversed. We showed earlier that when $12 \times 42 = 504$, and if we reverse the digits of each of the two numbers, we get $21 \times 24 = 504$. The same is true for the number pair 36 and 84 because $36 \times 84 = 3,024 = 63 \times 48$.

At this point you may wonder if this will happen with any pair of numbers. The answer is that it will only work with the following 14 pairs of numbers:

| | |
|---|---|
| $12 \times 42 = 21 \times 24 = 504$ | $12 \times 63 = 21 \times 36 = 756$ |
| $12 \times 84 = 21 \times 48 = 1008$ | $13 \times 62 = 31 \times 26 = 806$ |
| $13 \times 93 = 31 \times 39 = 1209$ | $14 \times 82 = 41 \times 28 = 1148$ |
| $23 \times 64 = 32 \times 46 = 1472$ | $23 \times 96 = 32 \times 69 = 2208$ |
| $24 \times 63 = 42 \times 36 = 1512$ | $24 \times 84 = 42 \times 48 = 2016$ |
| $26 \times 93 = 62 \times 39 = 2418$ | $34 \times 86 = 43 \times 68 = 2924$ |
| $36 \times 84 = 63 \times 48 = 3024$ | $46 \times 96 = 64 \times 69 = 4416$ |

**Equation 35**

A careful inspection of these 14 pairs of numbers will reveal that in each case the product of the tens digits of each pair of numbers is equal to the product of the units digits. We can justify this algebraically as follows: For the numbers $z_1$, $z_2$, $z_3$, and $z_4$, we have:

$$z_1 z_2 = (10a + b) \times (10c + d) = 100ac + 10ad + 10bc + bd, \text{ and}$$

$$z_3 z_4 = (10b + a) \times (10d + c) = 100bd + 10bc + 10ad + ac.$$

Here $a$, $b$, $c$, $d$ represent any of the ten digits: 0, 1, 2, ..., 9, where $a \neq 0$ and $c \neq 0$. We would like to show that $z_1 z_2 = z_3 z_4$. Therefore, $100ac + 10ad + 10bc + bd = 100bd + 10bc + 10ad + ac$, then $100ac + bd = 100bd + ac$, and $99ac = 99bd$, or $ac = bd$, which is what we observed earlier.

## SUMS OF CONSECUTIVE INTEGERS

Which numbers can be expressed as the sum of consecutive integers? You may want to experiment a bit before trying to generate a rule. Try to express the first batch of natural numbers as the sum of consecutive integers. We will provide some in the list in Figure 77.

| | |
|---|---|
| 2 = not possible | 21 = 1 + 2 + 3 + 4 + 5 + 6 |
| 3 = 1 + 2 | 22 = 4 + 5 + 6 + 7 |
| 4 = not possible | 23 = 11 + 12 |
| 5 = 2 + 3 | 24 = 7 + 8 + 9 |
| 6 = 1 + 2 + 3 | 25 = 12 + 13 |
| 7 = 3 + 4 | 26 = 5 + 6 + 7 + 8 |
| 8 = not possible | 27 = 8 + 9 + 10 |
| 9 = 4 + 5 | 28 = 1 + 2 + 3 + 4 + 5 + 6 + 7 |
| 10 = 1 + 2 + 3 + 4 | 29 = 14 + 15 |
| 11 = 5 + 6 | 30 = 4 + 5 + 6 + 7 + 8 |
| 12 = 3 + 4 + 5 | 31 = 15 + 16 |
| 13 = 6 + 7 | 32 = not possible |
| 14 = 2 + 3 + 4 + 5 | 33 = 10 + 11 + 12 |
| 15 = 4 + 5 + 6 | 34 = 7 + 8 + 9 + 10 |
| 16 = not possible | 35 = 17 + 18 |
| 17 = 8 + 9 | 36 = 1 + 2 + 3 + 4 + 5 + 6 + 7 + 8 |
| 18 = 5 + 6 + 7 | 37 = 18 + 19 |
| 19 = 9 + 10 | 38 = 8 + 9 + 10 + 11 |
| 20 = 2 + 3 + 4 + 5 + 6 | 39 = 19 + 20 |
| | 40 = 6 + 7 + 8 + 9 + 10 |

**Figure 77**

These consecutive-number-sum representations are clearly not unique. For example, 30 can be expressed in more than one way: $30 = 9 + 10 + 11$, or $30 = 6 + 7 + 8 + 9$.

An inspection of the table shows that those where a consecutive number sum was not possible are the powers of 2. This is an interesting fact. It is not something that one would expect. By making a list of these consecutive number sums, you might begin to see patterns. Clearly the triangular numbers are equal to the sum of the first $n$ natural numbers.[7] A multiple of 3, say $3n$, can always be represented by the sum: $(n - 1) + n + (n + 1)$. You may discover other patterns. That's part of the fun of it—not to mention the instructional value of seeing number patterns and relationships.

For the purists and the more ambitious reader, we now prove this (until now) conjecture. First, we will establish when a number can be expressed as a sum of at least two positive integers.

Let us analyze what values can be taken by the sum (S) of (two or more) consecutive positive integers from $a$ to $b$, where $b > a$. By applying the formula for the sum of an arithmetic series[8] we get:

$$S = a + (a + 1) + (a + 2) + \ldots + (b + 1) + b = \left(\frac{a + b}{2}\right)(b - a + 1).$$

Then doubling both sides of the equation, we get: $2S = (a + b)(b - a + 1)$.

Letting $(a + b) = x$ and $(b - a + 1) = y$, we can note that $x$ and $y$ are both integers, and that because their sum, $x + y = 2b + 1$, is odd, one of $x$, $y$ is odd and the other is even. Note that $2S = xy$.

**Case 1**. $S$ is a power of 2.

Let $S = 2^n$. We have $2(2^n) = xy$, or $2^{n+1} = xy$. The only way we can express $2^{n+1}$ as a product of an even and an odd number is if the odd number is 1. If $x = a + b = 1$, then $a$ and $b$ cannot be positive integers. If $y = b - a + 1 = 1$, then we have $a = b$, which also cannot occur. Therefore, $S$ cannot be a power of 2.

**Case 2**. $S$ is not a power of 2.

Let $S = m2^n$, where $m$ is an odd number greater than 1. We have $2(m2^n) = xy$, or $m2^{n+1} = xy$. We will now find positive integers $a$ and $b$ such that $b > a$ and $S = a + (a + 1) + \ldots + b$.

The two numbers $2^{n+1}$ and $m$ are not equal because one is odd and the other is even. Therefore, one is bigger than the other. Assign $x$ to be the bigger one and $y$ to be the smaller one. This assignment gives us a solution for $a$ and $b$, as $x + y = 2b + 1$, giving a positive integer value for $b$, and $x - y = 2a - 1$, giving a positive integer value for $a$. Also, $y = b - a + 1 > 1$, so $b > a$, as required. We have obtained $a$ and $b$.

Therefore, for any $S$ that is not a power of 2 we can find positive integers $a$ and $b$, $b > a$, such that $S = a + (a + 1) + \ldots + b$.

Just to recap, a number can be expressed as a sum of (at least two) positive integers, if and only if, the number is not a power of 2. This may appear harmless, but you should know that we have just accomplished some nice mathematics, while admiring the beauty of the subject.

## AMAZING POWER RELATIONSHIPS

Although we covered some of the topics presented in this chapter, it's still bears emphasis as a separate topic. Our number system has many unusual features built into it. Discovering them can certainly be a rewarding experience. Sometimes we stumble onto these relationship and other times they are the result of experimentation and avid searching—based on a hunch. The famous mathematician Carl Friedrich Gauss (1777–1855) discovered a fair number of relationships (which he later proved to establish theorems) on the basis of his superior arithmetic abilities.

Consider the following relationship and describe what is going on here.

$$81 = (8 + 1)^2 = 9^2$$

We have taken the square of the sum of the digits and again as follows:

$$4,913 = (4 + 9 + 1 + 3)^3 = 17^3$$

In both cases, we have taken the sum of the digits to a power and ended up with the number we started with. Impressed? You ought to be, for this is quite astonishing. Now, to find other such numbers is no mean feat.

The list in Figure 78 will provide you with lots of examples of these unusual numbers. Enjoy yourself!

The beauty is self-evident!

| Number | = | (Sum of the Digits)$^n$ | Number | = | (Sum of the Digits)$^n$ |
|---|---|---|---|---|---|
| 81 | = | $9^2$ | 34,012,224 | = | $18^6$ |
|  |  |  | 8,303,765,625 | = | $45^6$ |
| 512 | = | $8^3$ | 24,794,911,296 | = | $54^6$ |
| 4,913 | = | $17^3$ | 68,719,476,736 | = | $64^6$ |
| 5,832 | = | $18^3$ |  |  |  |
| 17,576 | = | $26^3$ | 612,220,032 | = | $18^7$ |
| 19,683 | = | $27^3$ | 10,460,353,203 | = | $27^7$ |
|  |  |  | 27,512,614,111 | = | $31^7$ |
| 2,401 | = | $7^4$ | 52,523,350,144 | = | $34^7$ |
| 234,256 | = | $22^4$ | 271,818,611,107 | = | $43^7$ |
| 390,625 | = | $25^4$ | 1,174,711,139,837 | = | $53^7$ |
| 614,656 | = | $28^4$ | 2,207,984,167,552 | = | $58^7$ |
| 1,679,616 | = | $36^4$ | 6,722,988,818,432 | = | $68^7$ |
|  |  |  |  |  |  |
| 17,210,368 | = | $28^5$ | 20,047,612,231,936 | = | $46^8$ |
| 52,521,875 | = | $35^5$ | 72,301,961,339,136 | = | $54^8$ |
| 60,466,176 | = | $36^5$ | 248,155,780,267,521 | = | $63^8$ |
| 205,962,976 | = | $46^5$ |  |  |  |

| | | |
|---|---|---|
| 20,864,448,472,975,628,947,226,005,981,267,194,447,042,584,001 | = | $207^{20}$ |

**Figure 78**

## A FACTORIAL LOOP

This little charming concept will show an unusual relationship for certain numbers that we have considered earlier but still ought to be considered for the process rather than for the individual numbers alone. But before we begin, we once again review the factorial notation. The definition of $n!$ is $n! = 1 \cdot 2 \cdot 3 \cdot 4 \ldots \ldots (n-1) \cdot n.$ [9]

When we find the sum of the factorials of the digits of 145, we get: $1! + 4! + 5! = 1 + 24 + 120 = 145$. Surprise! We are back to 145. Only for certain numbers, will the sum of the factorials of the digits equal the number itself, as we have seen through our journey of the numbers earlier.

Let's try this again with the number 40,585. That is, $4! + 0! + 5! + 8! + 5! = 24 + 1 + 120 + 40,320 + 120 = 40,585$. It is natural that you would expect this to be true for just about any number. Well, just try another number. Chances are that it will not work. Suppose we begin with the number 871.

Using this procedure, you will get: $8! + 7! + 1! = 40,320 + 5,040 + 1 = 45,361$, at this point you may feel that you have failed. Not so fast. Try this procedure again with the previous result 45,361.

This yields: $4! + 5! + 3! + 6! + 1! = 24 + 120 + 6 + 720 + 1 = 871$. Isn't this the very number we started with? Again, we formed a loop.

If you repeat this with the number 872, you will get $8! + 7! + 2! = 40,320 + 5,040 + 2 = 45,361$. Then repeating the process will give you: $4! + 5! + 3! + 6! + 2! = 24 + 120 + 6 + 720 + 2 = 872$. Again, we're in a loop.

Some people are usually quick to form generalizations, so they might conclude that if the scheme of summing factorials of the digits of a number doesn't get you back to the original number then try it again and it ought to work. Of course, you can "stack the deck" by considering the number 169. Two cycles do not seem to present a loop. So, proceed through one more cycle. And sure enough, the third cycle leads back to the original number (see Figure 79).

| Starting number | Sum of the factorials |
|---|---|
| 169 | $1! + 6! + 9! = 363,601$ |
| 363,601 | $3! + 6! + 3! + 6! + 0! + 1! =$ <br> $6 + 720 + 6 + 720 + 1 + 1 = 1,454$ |
| 1,454 | $1! + 4! + 5! + 4! =$ <br> $1 + 24 + 120 + 24 = 169$ |

Figure 79

Be careful about drawing conclusions. These factorial oddities are not so pervasive that you should try to find others. There are "within reach" three groups of such loops. We can organize them according to the number of times you have to repeat the process to reach the original number. We will call these repetitions "cycles." Here is a summary of the way our numbers behave in this factorial loop.

| 1 cycle | 1, 2, 145, 40,585 |
|---------|-------------------|
| 2 cycle | 871, 45361 and 872, 45362 |
| 3 cycle | 169, 363601, 1454 |

**Equation 36**

The factorial loops shown in this charming little number oddity can be fun, but you ought to be cautioned that there are no other such numbers less than 2,000,000 for which this works. So, don't waste your time in a fruitless search. Just appreciate some little beauties!

## NOTES

1. $M_p = 2^p - 1$ and $p$ is prime.
2. Working backward.
3. Repunits are defined as $r_n = \dfrac{10^n - 1}{9}$, where $n \in \mathbb{N}$. In these numbers the units digit 1 repeats $n$ times and are therefore, called *repunits* (repeated units).
4. Richard L. Francis, "Mathematical Haystacks: Another Look at Repunit Numbers," *The College Mathematics Journal* 19(3) (May 1988): 240–46.
5. http://mathworld.wolfram.com/Repunit.html
6. For a proof that this relationship holds as started, see Ross Honsberger, *Ingenuity in Mathematics*. New York: Random House, 1970, 147–56.
7. Remember that the natural numbers are the counting numbers: 1, 2, 3, 4, 5, 6, . . . .
8. The sum of an arithmetic series is $S = \dfrac{n}{2}(a + l)$, where $n$ is the number of terms, and $a$ is the first term and $l$ is the last term.
9. The factorial notation "!" represents the product of all the integers equal to and less than the number with the factorial notation. For example, $5! = (5)(4)(3)(2)(1) = 120$, $8! = (8)(7)(6)(5)(4)(3)(2)(1) = 40{,}320$, and $n! = (n)(n-1)(n-2)(n-3) \ldots (3)(2)(1)$. Notice here we are using the parentheses to indicate multiplication.

# Epilogue

We have now investigated numbers of all kinds and the many amazing and unexpected relationships that they harbor. They relate to each other and two special arrangements the likes of which most people are unaware. By investigating many of these incredible relationships the reader will have gotten a much deeper insight into the nature of numbers and the secrets that they hold. Once again, the reader will have been exposed to the beauty and power that mathematics encompasses, and that unfortunately all too often is overseen in the normal traditional study of the subject. With this book we hope to have motivated the reader to pursue further investigations.

# Appendix: Tables

## 1. Table of Fibonacci numbers

| No. | Fibonacci | No. | Fibonacci | No. | Fibonacci | No. | Fibonacci |
|---|---|---|---|---|---|---|---|
| 1 | 1 | 21 | 10946 | 41 | 165580141 | 61 | 2504730781961 |
| 2 | 1 | 22 | 17711 | 42 | 267914296 | 62 | 4052739537881 |
| 3 | 2 | 23 | 28657 | 43 | 433494437 | 63 | 6557470319842 |
| 4 | 3 | 24 | 46368 | 44 | 701408733 | 64 | 10610209857723 |
| 5 | 5 | 25 | 75025 | 45 | 1134903170 | 65 | 17167680177565 |
| 6 | 8 | 26 | 121393 | 46 | 1836311903 | 66 | 27777890035288 |
| 7 | 13 | 27 | 196418 | 47 | 2971215073 | 67 | 44945570212853 |
| 8 | 21 | 28 | 317811 | 48 | 4807526976 | 68 | 72723460248141 |
| 9 | 34 | 29 | 514229 | 49 | 7778742049 | 69 | 117669030460994 |
| 10 | 55 | 30 | 832040 | 50 | 12586269025 | 70 | 190392490709135 |
| 11 | 89 | 31 | 1346269 | 51 | 20365011074 | 71 | 308061521170129 |
| 12 | 144 | 32 | 2178309 | 52 | 32951280099 | 72 | 498454011879264 |
| 13 | 233 | 33 | 3524578 | 53 | 53316291173 | 73 | 806515533049393 |
| 14 | 377 | 34 | 5702887 | 54 | 86267571272 | 74 | 1304969544928657 |
| 15 | 610 | 35 | 9227465 | 55 | 139583862445 | 75 | 2111485077978050 |
| 16 | 987 | 36 | 14930352 | 56 | 225851433717 | 76 | 3416454622906707 |
| 17 | 1597 | 37 | 24157817 | 57 | 365435296162 | 77 | 5527939700884757 |
| 18 | 2584 | 38 | 39088169 | 58 | 591286729879 | 78 | 8944394323791464 |
| 19 | 4181 | 39 | 63245986 | 59 | 956722026041 | 79 | 14472334024676221 |
| 20 | 6765 | 40 | 102334155 | 60 | 1548008755920 | 80 | 23416728348467685 |

| No. | Fibonacci | No. | Fibonacci |
|---|---|---|---|
| 81 | 37889062373143906 | 96 | 51680708854858323072 |
| 82 | 61305790721611591 | 97 | 83621143489848422977 |
| 83 | 99194853094755497 | 98 | 135301852344706746049 |
| 84 | 160500643816367088 | 99 | 218922995834555169026 |
| 85 | 259695496911122585 | 100 | 354224848179261915075 |
| 86 | 420196140727489673 | 101 | 573147844013817084101 |
| 87 | 679891637638612258 | 102 | 927372692193078999176 |
| 88 | 1100087778366101931 | 103 | 1500520536206896083277 |
| 89 | 1779979416004714189 | 104 | 2427893228399975082453 |
| 90 | 2880067194370816120 | 105 | 3928413764606871165730 |
| 91 | 4660046610375530309 | 106 | 6356306993006846248183 |
| 92 | 7540113804746346429 | 107 | 10284720757613717413913 |
| 93 | 12200160415121876738 | 108 | 16641027750620563662096 |
| 94 | 19740274219868223167 | 109 | 26925748508234281076009 |
| 95 | 31940434634990099905 | 110 | 43566776258854844738105 |

**Figure D.1**

## 2. Table of the prime numbers less than 10,000

| 2 | 3 | 5 | 7 | 11 | 13 | 17 | 19 | 23 | 29 | 31 | 37 |
|---|---|---|---|----|----|----|----|----|----|----|----|
| 41 | 43 | 47 | 53 | 59 | 61 | 67 | 71 | 73 | 79 | 83 | 89 |
| 97 | 101 | 103 | 107 | 109 | 113 | 127 | 131 | 137 | 139 | 149 | 151 |
| 157 | 163 | 167 | 173 | 179 | 181 | 191 | 193 | 197 | 199 | 211 | 223 |
| 227 | 229 | 233 | 239 | 241 | 251 | 257 | 263 | 269 | 271 | 277 | 281 |
| 283 | 293 | 307 | 311 | 313 | 317 | 331 | 337 | 347 | 349 | 353 | 359 |
| 367 | 373 | 379 | 383 | 389 | 397 | 401 | 409 | 419 | 421 | 431 | 433 |
| 439 | 443 | 449 | 457 | 461 | 463 | 467 | 479 | 487 | 491 | 499 | 503 |
| 509 | 521 | 523 | 541 | 547 | 557 | 563 | 569 | 571 | 577 | 587 | 593 |
| 599 | 601 | 607 | 613 | 617 | 619 | 631 | 641 | 643 | 647 | 653 | 659 |
| 661 | 673 | 677 | 683 | 691 | 701 | 709 | 719 | 727 | 733 | 739 | 743 |
| 751 | 757 | 761 | 769 | 773 | 787 | 797 | 809 | 811 | 821 | 823 | 827 |
| 829 | 839 | 853 | 857 | 859 | 863 | 877 | 881 | 883 | 887 | 907 | 911 |
| 919 | 929 | 937 | 941 | 947 | 953 | 967 | 971 | 977 | 983 | 991 | 997 |
| 1009 | 1013 | 1019 | 1021 | 1031 | 1033 | 1039 | 1049 | 1051 | 1061 | 1063 | 1069 |
| 1087 | 1091 | 1093 | 1097 | 1103 | 1109 | 1117 | 1123 | 1129 | 1151 | 1153 | 1163 |
| 1171 | 1181 | 1187 | 1193 | 1201 | 1213 | 1217 | 1223 | 1229 | 1231 | 1237 | 1249 |
| 1259 | 1277 | 1279 | 1283 | 1289 | 1291 | 1297 | 1301 | 1303 | 1307 | 1319 | 1321 |
| 1327 | 1361 | 1367 | 1373 | 1381 | 1399 | 1409 | 1423 | 1427 | 1429 | 1433 | 1439 |
| 1447 | 1451 | 1453 | 1459 | 1471 | 1481 | 1483 | 1487 | 1489 | 1493 | 1499 | 1511 |
| 1523 | 1531 | 1543 | 1549 | 1553 | 1559 | 1567 | 1571 | 1579 | 1583 | 1597 | 1601 |
| 1607 | 1609 | 1613 | 1619 | 1621 | 1627 | 1637 | 1657 | 1663 | 1667 | 1669 | 1693 |
| 1697 | 1699 | 1709 | 1721 | 1723 | 1733 | 1741 | 1747 | 1753 | 1759 | 1777 | 1783 |
| 1787 | 1789 | 1801 | 1811 | 1823 | 1831 | 1847 | 1861 | 1867 | 1871 | 1873 | 1877 |
| 1879 | 1889 | 1901 | 1907 | 1913 | 1931 | 1933 | 1949 | 1951 | 1973 | 1979 | 1987 |
| 1993 | 1997 | 1999 | 2003 | 2011 | 2017 | 2027 | 2029 | 2039 | 2053 | 2063 | 2069 |
| 2081 | 2083 | 2087 | 2089 | 2099 | 2111 | 2113 | 2129 | 2131 | 2137 | 2141 | 2143 |
| 2153 | 2161 | 2179 | 2203 | 2207 | 2213 | 2221 | 2237 | 2239 | 2243 | 2251 | 2267 |
| 2269 | 2273 | 2281 | 2287 | 2293 | 2297 | 2309 | 2311 | 2333 | 2339 | 2341 | 2347 |
| 2351 | 2357 | 2371 | 2377 | 2381 | 2383 | 2389 | 2393 | 2399 | 2411 | 2417 | 2423 |
| 2437 | 2441 | 2447 | 2459 | 2467 | 2473 | 2477 | 2503 | 2521 | 2531 | 2539 | 2543 |
| 2549 | 2551 | 2557 | 2579 | 2591 | 2593 | 2609 | 2617 | 2621 | 2633 | 2647 | 2657 |
| 2659 | 2663 | 2671 | 2677 | 2683 | 2687 | 2689 | 2693 | 2699 | 2707 | 2711 | 2713 |
| 2719 | 2729 | 2731 | 2741 | 2749 | 2753 | 2767 | 2777 | 2789 | 2791 | 2797 | 2801 |
| 2803 | 2819 | 2833 | 2837 | 2843 | 2851 | 2857 | 2861 | 2879 | 2887 | 2897 | 2903 |
| 2909 | 2917 | 2927 | 2939 | 2953 | 2957 | 2963 | 2969 | 2971 | 2999 | 3001 | 3011 |
| 3019 | 3023 | 3037 | 3041 | 3049 | 3061 | 3067 | 3079 | 3083 | 3089 | 3109 | 3119 |
| 3121 | 3137 | 3163 | 3167 | 3169 | 3181 | 3187 | 3191 | 3203 | 3209 | 3217 | 3221 |
| 3229 | 3251 | 3253 | 3257 | 3259 | 3271 | 3299 | 3301 | 3307 | 3313 | 3319 | 3323 |
| 3329 | 3331 | 3343 | 3347 | 3359 | 3361 | 3371 | 3373 | 3389 | 3391 | 3407 | 3413 |
| 3433 | 3449 | 3457 | 3461 | 3463 | 3467 | 3469 | 3491 | 3499 | 3511 | 3517 | 3527 |
| 3529 | 3533 | 3539 | 3541 | 3547 | 3557 | 3559 | 3571 | 3581 | 3583 | 3593 | 3607 |
| 3613 | 3617 | 3623 | 3631 | 3637 | 3643 | 3659 | 3671 | 3673 | 3677 | 3691 | 3697 |

**Figure D.2**

| | | | | | | | | | | | |
|---|---|---|---|---|---|---|---|---|---|---|---|
| 3701 | 3709 | 3719 | 3727 | 3733 | 3739 | 3761 | 3767 | 3769 | 3779 | 3793 | 3797 |
| 3803 | 3821 | 3823 | 3833 | 3847 | 3851 | 3853 | 3863 | 3877 | 3881 | 3889 | 3907 |
| 3911 | 3917 | 3919 | 3923 | 3929 | 3931 | 3943 | 3947 | 3967 | 3989 | 4001 | 4003 |
| 4007 | 4013 | 4019 | 4021 | 4027 | 4049 | 4051 | 4057 | 4073 | 4079 | 4091 | 4093 |
| 4099 | 4111 | 4127 | 4129 | 4133 | 4139 | 4153 | 4157 | 4159 | 4177 | 4201 | 4211 |
| 4217 | 4219 | 4229 | 4231 | 4241 | 4243 | 4253 | 4259 | 4261 | 4271 | 4273 | 4283 |
| 4289 | 4297 | 4327 | 4337 | 4339 | 4349 | 4357 | 4363 | 4373 | 4391 | 4397 | 4409 |
| 4421 | 4423 | 4441 | 4447 | 4451 | 4457 | 4463 | 4481 | 4483 | 4493 | 4507 | 4513 |
| 4517 | 4519 | 4523 | 4547 | 4549 | 4561 | 4567 | 4583 | 4591 | 4597 | 4603 | 4621 |
| 4637 | 4639 | 4643 | 4649 | 4651 | 4657 | 4663 | 4673 | 4679 | 4691 | 4703 | 4721 |
| 4723 | 4729 | 4733 | 4751 | 4759 | 4783 | 4787 | 4789 | 4793 | 4799 | 4801 | 4813 |
| 4817 | 4831 | 4861 | 4871 | 4877 | 4889 | 4903 | 4909 | 4919 | 4931 | 4933 | 4937 |
| 4943 | 4951 | 4957 | 4967 | 4969 | 4973 | 4987 | 4993 | 4999 | 5003 | 5009 | 5011 |
| 5021 | 5023 | 5039 | 5051 | 5059 | 5077 | 5081 | 5087 | 5099 | 5101 | 5107 | 5113 |
| 5119 | 5147 | 5153 | 5167 | 5171 | 5179 | 5189 | 5197 | 5209 | 5227 | 5231 | 5233 |
| 5237 | 5261 | 5273 | 5279 | 5281 | 5297 | 5303 | 5309 | 5323 | 5333 | 5347 | 5351 |
| 5381 | 5387 | 5393 | 5399 | 5407 | 5413 | 5417 | 5419 | 5431 | 5437 | 5441 | 5443 |
| 5449 | 5471 | 5477 | 5479 | 5483 | 5501 | 5503 | 5507 | 5519 | 5521 | 5527 | 5531 |
| 5557 | 5563 | 5569 | 5573 | 5581 | 5591 | 5623 | 5639 | 5641 | 5647 | 5651 | 5653 |
| 5657 | 5659 | 5669 | 5683 | 5689 | 5693 | 5701 | 5711 | 5717 | 5737 | 5741 | 5743 |
| 5749 | 5779 | 5783 | 5791 | 5801 | 5807 | 5813 | 5821 | 5827 | 5839 | 5843 | 5849 |
| 5851 | 5857 | 5861 | 5867 | 5869 | 5879 | 5881 | 5897 | 5903 | 5923 | 5927 | 5939 |
| 5953 | 5981 | 5987 | 6007 | 6011 | 6029 | 6037 | 6043 | 6047 | 6053 | 6067 | 6073 |
| 6079 | 6089 | 6091 | 6101 | 6113 | 6121 | 6131 | 6133 | 6143 | 6151 | 6163 | 6173 |
| 6197 | 6199 | 6203 | 6211 | 6217 | 6221 | 6229 | 6247 | 6257 | 6263 | 6269 | 6271 |
| 6277 | 6287 | 6299 | 6301 | 6311 | 6317 | 6323 | 6329 | 6337 | 6343 | 6353 | 6359 |
| 6361 | 6367 | 6373 | 6379 | 6389 | 6397 | 6421 | 6427 | 6449 | 6451 | 6469 | 6473 |
| 6481 | 6491 | 6521 | 6529 | 6547 | 6551 | 6553 | 6563 | 6569 | 6571 | 6577 | 6581 |
| 6599 | 6607 | 6619 | 6637 | 6653 | 6659 | 6661 | 6673 | 6679 | 6689 | 6691 | 6701 |
| 6703 | 6709 | 6719 | 6733 | 6737 | 6761 | 6763 | 6779 | 6781 | 6791 | 6793 | 6803 |
| 6823 | 6827 | 6829 | 6833 | 6841 | 6857 | 6863 | 6869 | 6871 | 6883 | 6899 | 6907 |
| 6911 | 6917 | 6947 | 6949 | 6959 | 6961 | 6967 | 6971 | 6977 | 6983 | 6991 | 6997 |
| 7001 | 7013 | 7019 | 7027 | 7039 | 7043 | 7057 | 7069 | 7079 | 7103 | 7109 | 7121 |
| 7127 | 7129 | 7151 | 7159 | 7177 | 7187 | 7193 | 7207 | 7211 | 7213 | 7219 | 7229 |
| 7237 | 7243 | 7247 | 7253 | 7283 | 7297 | 7307 | 7309 | 7321 | 7331 | 7333 | 7349 |
| 7351 | 7369 | 7393 | 7411 | 7417 | 7433 | 7451 | 7457 | 7459 | 7477 | 7481 | 7487 |
| 7489 | 7499 | 7507 | 7517 | 7523 | 7529 | 7537 | 7541 | 7547 | 7549 | 7559 | 7561 |
| 7573 | 7577 | 7583 | 7589 | 7591 | 7603 | 7607 | 7621 | 7639 | 7643 | 7649 | 7669 |
| 7673 | 7681 | 7687 | 7691 | 7699 | 7703 | 7717 | 7723 | 7727 | 7741 | 7753 | 7757 |
| 7759 | 7789 | 7793 | 7817 | 7823 | 7829 | 7841 | 7853 | 7867 | 7873 | 7877 | 7879 |
| 7883 | 7901 | 7907 | 7919 | 7927 | 7933 | 7937 | 7949 | 7951 | 7963 | 7993 | 8009 |
| 8011 | 8017 | 8039 | 8053 | 8059 | 8069 | 8081 | 8087 | 8089 | 8093 | 8101 | 8111 |
| 8117 | 8123 | 8147 | 8161 | 8167 | 8171 | 8179 | 8191 | 8209 | 8219 | 8221 | 8231 |

**Figure D.2** *(continued)*

| 8233 | 8237 | 8243 | 8263 | 8269 | 8273 | 8287 | 8291 | 8293 | 8297 | 8311 | 8317 |
| 8329 | 8353 | 8363 | 8369 | 8377 | 8387 | 8389 | 8419 | 8423 | 8429 | 8431 | 8443 |
| 8447 | 8461 | 8467 | 8501 | 8513 | 8521 | 8527 | 8537 | 8539 | 8543 | 8563 | 8573 |
| 8581 | 8597 | 8599 | 8609 | 8623 | 8627 | 8629 | 8641 | 8647 | 8663 | 8669 | 8677 |
| 8681 | 8689 | 8693 | 8699 | 8707 | 8713 | 8719 | 8731 | 8737 | 8741 | 8747 | 8753 |
| 8761 | 8779 | 8783 | 8803 | 8807 | 8819 | 8821 | 8831 | 8837 | 8839 | 8849 | 8861 |
| 8863 | 8867 | 8887 | 8893 | 8923 | 8929 | 8933 | 8941 | 8951 | 8963 | 8969 | 8971 |
| 8999 | 9001 | 9007 | 9011 | 9013 | 9029 | 9041 | 9043 | 9049 | 9059 | 9067 | 9091 |
| 9103 | 9109 | 9127 | 9133 | 9137 | 9151 | 9157 | 9161 | 9173 | 9181 | 9187 | 9199 |
| 9203 | 9209 | 9221 | 9227 | 9239 | 9241 | 9257 | 9277 | 9281 | 9283 | 9293 | 9311 |
| 9319 | 9323 | 9337 | 9341 | 9343 | 9349 | 9371 | 9377 | 9391 | 9397 | 9403 | 9413 |
| 9419 | 9421 | 9431 | 9433 | 9437 | 9439 | 9461 | 9463 | 9467 | 9473 | 9479 | 9491 |
| 9497 | 9511 | 9521 | 9533 | 9539 | 9547 | 9551 | 9587 | 9601 | 9613 | 9619 | 9623 |
| 9629 | 9631 | 9643 | 9649 | 9661 | 9677 | 9679 | 9689 | 9697 | 9719 | 9721 | 9733 |
| 9739 | 9743 | 9749 | 9767 | 9769 | 9781 | 9787 | 9791 | 9803 | 9811 | 9817 | 9829 |
| 9833 | 9839 | 9851 | 9857 | 9859 | 9871 | 9883 | 9887 | 9901 | 9907 | 9923 | 9929 |
| 9931 | 9941 | 9949 | 9967 | 9973 | | | | | | | |

**Figure D.2**

## 3.  Table of the first 48 perfect numbers

| k | Perfect number | Number of Digits | Year discovered |
|---|---|---|---|
| 2 | 6 | 1 | antiquity |
| 3 | 28 | 2 | antiquity |
| 5 | 496 | 3 | antiquity |
| 7 | 8128 | 4 | antiquity |
| 13 | 33550336 | 8 | 1456 |
| 17 | 8589869056 | 10 | 1588 |
| 19 | 137438691328 | 12 | 1588 |
| 31 | 2305843008139952128 | 19 | 1772 |
| 61 | 265845599...953842176 | 37 | 1883 |
| 89 | 191561942...548169216 | 54 | 1911 |
| 107 | 131640364...783728128 | 65 | 1914 |
| 127 | 144740111...199152128 | 77 | 1876 |
| 521 | 235627234...555646976 | 314 | 1952 |
| 607 | 141053783...537328128 | 366 | 1952 |
| 1279 | 541625262...984291328 | 770 | 1952 |
| 2203 | 108925835...453782528 | 1327 | 1952 |
| 2281 | 994970543...139915776 | 1373 | 1952 |
| 3217 | 335708321...628525056 | 1937 | 1957 |
| 4253 | 182017490...133377536 | 2561 | 1961 |
| 4423 | 407672717...912534528 | 2663 | 1961 |
| 9689 | 114347317...429577216 | 5834 | 1963 |
| 9941 | 598885496...073496576 | 5985 | 1963 |
| 11213 | 395961321...691086336 | 6751 | 1963 |
| 19937 | 931144559...271942656 | 12003 | 1971 |
| 21701 | 100656497...141605376 | 13066 | 1978 |
| 23209 | 811537765...941666816 | 13973 | 1979 |
| 44497 | 365093519...031827456 | 26790 | 1979 |
| 86243 | 144145836...360406528 | 51924 | 1982 |
| 110503 | 136204582...603862528 | 66530 | 1988 |
| 132049 | 131451295...774550016 | 79502 | 1983 |
| 216091 | 278327459...840880128 | 130100 | 1985 |
| 756839 | 151616570...565731328 | 455663 | 1992 |
| 859433 | 838488226...416167936 | 517430 | 1994 |

**Figure D.3** *(continued)*

| 1257787 | 849732889...118704128 | 757263 | 1996 |
|---------|----------------------|--------|------|
| 1398269 | 331882354...723375616 | 841842 | 1996 |
| 2976221 | 194276425...174462976 | 1791864 | 1997 |
| 3021377 | 811686848...022457856 | 1819050 | 1998 |
| 6972593 | 955176030...123572736 | 4197919 | 1999 |
| 13466917 | 427764159...863021056 | 8107892 | 2001 |
| 20996011 | 793508909...206896128 | 12640858 | 2003 |
| 24036583 | 448233026...572950528 | 14471465 | 2004 |
| 25964951 | 746209841...791088128 | 15632458 | 2005 |
| 30402457 | 497437765...164704256 | 18304103 | 2005 |
| 32582657 | 775946855...577120256 | 19616714 | 2006 |
| 37156667 | 204534225...074480128 | 22370543 | 2008 |
| 42643801 | 144285057...377253376 | 25674127 | 2009 |
| 43112609 | 500767156...145378816 | 25956377 | 2008 |
| 57885161 | 169296395...270130176 | 34850340 | 2013 |

**Figure D.3**

## 4. Table of Kaprekar numbers

| Kaprekar number | Square of the number | | Decomposition |
|---|---|---|---|
| 1 | $1^2 =$ | 1 | $1 = 1$ |
| 9 | $9^2 =$ | 81 | $8 + 1 = 9$ |
| 45 | $45^2 =$ | 2,025 | $20 + 25 = 45$ |
| 55 | $55^2 =$ | 3,025 | $30 + 25 = 55$ |
| 99 | $99^2 =$ | 9,801 | $98 + 01 = 99$ |
| 297 | $297^2 =$ | 88,209 | $88 + 209 = 297$ |
| 703 | $703^2 =$ | 494,209 | $494 + 209 = 703$ |
| 999 | $999^2 =$ | 998,001 | $998 + 001 = 999$ |
| 2,223 | $2,223^2 =$ | 4,941,729 | $494 + 1,729 = 2,223$ |
| 2,728 | $2728^2 =$ | 7,441,984 | $744 + 1,984 = 2,728$ |
| 4,879 | $4,879^2 =$ | 23,804,641 | $238 + 04,641 = 4,879$ |
| 4,950 | $4,950^2 =$ | 24,502,500 | $2,450 + 2,500 = 4,950$ |
| 5,050 | $5,050^2 =$ | 25,502,500 | $2,550 + 2,500 = 5,050$ |
| 5,292 | $5,292^2 =$ | 28,005,264 | $28 + 005,264 = 5,292$ |
| 7,272 | $7,272^2 =$ | 52,881,984 | $5,288 + 1,984 = 7,272$ |
| 7,777 | $7,777^2 =$ | 60,481,729 | $6,048 + 1,729 = 7,777$ |
| 9,999 | $999^2 =$ | 99,980,001 | $9,998 + 0,001 = 9,999$ |
| 17,344 | $17,344^2 =$ | 300,814,336 | $3,008 + 14,336 = 17,344$ |
| 22,222 | $22,222^2 =$ | 493,817,284 | $4,938 + 17,284 = 22,222$ |
| 38,962 | $38,962^2 =$ | 1,518,037,444 | $1,518 + 037,444 = 38,962$ |
| 77,778 | $77,778^2 =$ | 6,049,417,284 | $60,494 + 17,284 = 77,778$ |
| 82,656 | $82,656^2 =$ | 6,832,014,336 | $68,320 + 14,336 = 82,656$ |
| 95,121 | $95,121^2 =$ | 9,048,004,641 | $90,480 + 04,641 = 95,121$ |
| 99,999 | $99,999^2 =$ | 9,999,800,001 | $99,998 + 000001 = 99.999$ |
| 142,857 | $142,857^2 =$ | 20408122449 | $20,408 + 122,449 = 142,857$ |
| 148,149 | $148,149^2 =$ | 21948126201 | $21,948 + 126,201 = 148,149$ |
| 181,819 | $181,819^2 =$ | 33058148761 | $33,058 + 148,761 = 181,819$ |
| 187,110 | $187,110^2 =$ | 35010152100 | $35,010 + 152,100 = 187,110$ |

**Figure D.4**

The next Kaprekar numbers are 208,495, 318,682, 329,967, 351,352, 356,643, 390,313, 461,539, 466,830, 499,500, 500,500, 533,170, 857,143, . . .

## 5. Table of Armstrong numbers

| no. | digits | Armstrong number | | no. | digits | Armstrong number |
|---|---|---|---|---|---|---|
| 0 | 1 | 0 | | 45 | 17 | 35641594208964132 |
| 1 | 1 | 1 | | 46 | 17 | 35875699062250035 |
| 2 | 1 | 2 | | 47 | 19 | 1517841543307505039 |
| 3 | 1 | 3 | | 48 | 19 | 3289582984443187032 |
| 4 | 1 | 4 | | 49 | 19 | 4498128791164624869 |
| 5 | 1 | 5 | | 50 | 19 | 4929273885928088826 |
| 6 | 1 | 6 | | 51 | 20 | 63105425988599693916 |
| 7 | 1 | 7 | | 52 | 21 | 128468643043731391252 |
| 8 | 1 | 8 | | 53 | 21 | 449177399146038697307 |
| 9 | 1 | 9 | | 54 | 23 | 21887696841122916288858 |
| 10 | 3 | 153 | | 55 | 23 | 27879694893054074471405 |
| 11 | 3 | 370 | | 56 | 23 | 27907865009977052567814 |
| 12 | 3 | 371 | | 57 | 23 | 28361281321319229463398 |
| 13 | 3 | 407 | | 58 | 23 | 35452590104031691935943 |
| 14 | 4 | 1634 | | 59 | 24 | 174088005938065293023722 |
| 15 | 4 | 8208 | | 60 | 24 | 188451485447897896036875 |
| 16 | 4 | 9474 | | 61 | 24 | 239313664430041569350093 |
| 17 | 5 | 54748 | | 62 | 25 | 1550475334214501539088894 |
| 18 | 5 | 92727 | | 63 | 25 | 1553242162893771850669378 |
| 19 | 5 | 93084 | | 64 | 25 | 3706907995955475988644380 |
| 20 | 6 | 548834 | | 65 | 25 | 3706907995955475988644381 |
| 21 | 7 | 1741725 | | 66 | 25 | 4422095118095899619457938 |
| 22 | 7 | 4210818 | | 67 | 27 | 121204998563613372405438066 |
| 23 | 7 | 9800817 | | 68 | 27 | 121270696006801314328439376 |
| 24 | 7 | 9926315 | | 69 | 27 | 128851796696487777842012787 |
| 25 | 8 | 24678050 | | 70 | 27 | 174650464499531377631639254 |
| 26 | 8 | 24678051 | | 71 | 27 | 177265453171792792366489765 |
| 27 | 8 | 88593477 | | 72 | 29 | 14607640612971980372614873089 |
| 28 | 9 | 146511208 | | 73 | 29 | 19008174136254279995012734740 |
| 29 | 9 | 472335975 | | 74 | 29 | 19008174136254279995012734741 |
| 30 | 9 | 534494836 | | 75 | 29 | 23866716435523975980390369295 |
| 31 | 9 | 912985153 | | 76 | 31 | 1145037275765491025924292050346 |
| 32 | 10 | 4679307774 | | 77 | 31 | 1927890457142960697580636236639 |
| 33 | 11 | 32164049650 | | 78 | 31 | 2309092682616190307509695338915 |
| 34 | 11 | 32164049651 | | 79 | 32 | 17333509997782249308725103962772 |
| 35 | 11 | 40028394225 | | 80 | 33 | 186709961001538790100634132976990 |
| 36 | 11 | 42678290603 | | 81 | 33 | 186709961001538790100634132976991 |
| 37 | 11 | 44708635679 | | 82 | 34 | 1122763285329372541592822900204593 |
| 38 | 11 | 49388550606 | | 83 | 35 | 12639369517103790328947807201478392 |
| 39 | 11 | 82693916578 | | 84 | 35 | 12679937780272278566303885594196922 |
| 40 | 11 | 94204591914 | | 85 | 37 | 1219167219625434121569735803609966019 |
| 41 | 14 | 28116440335967 | | 86 | 38 | 12815792078366059955099770545296129367 |
| 42 | 16 | 4338281769391370 | | 87 | 39 | 115132219018763992565095597973971522400 |
| 43 | 16 | 4338281769391371 | | 88 | 39 | 115132219018763992565095597973971522401 |
| 44 | 17 | 21897142587612075 | | | | |

**Figure D.5**

## 6. Table of Amicable numbers

|    | First Number | Second Number | Year of discovery |
|----|--------------|---------------|-------------------|
| 1  | 220          | 284           | ca. 500 BCE-      |
| 2  | 1184         | 1210          | 1860              |
| 3  | 2620         | 2924          | 1747              |
| 4  | 5020         | 5564          | 1747              |
| 5  | 6232         | 6368          | 1747              |
| 6  | 10744        | 10856         | 1747              |
| 7  | 12285        | 14595         | 1939              |
| 8  | 17296        | 18416         | ca. 1310/1636     |
| 9  | 63020        | 76084         | 1747              |
| 10 | 66928        | 66992         | 1747              |
| 11 | 67095        | 71145         | 1747              |
| 12 | 69615        | 87633         | 1747              |
| 13 | 79750        | 88730         | 1964              |
| 14 | 100485       | 124155        | 1747              |
| 15 | 122265       | 139815        | 1747              |
| 16 | 122368       | 123152        | 1941/42           |
| 17 | 141664       | 153176        | 1747              |
| 18 | 142310       | 168730        | 1747              |
| 19 | 171856       | 176336        | 1747              |
| 20 | 176272       | 180848        | 1747              |
| 21 | 185368       | 203432        | 1966              |
| 22 | 196724       | 202444        | 1747              |
| 23 | 280540       | 365084        | 1966              |
| 24 | 308620       | 389924        | 1747              |
| 25 | 319550       | 430402        | 1966              |
| 26 | 356408       | 399592        | 1921              |
| 27 | 437456       | 455344        | 1747              |
| 28 | 469028       | 486178        | 1966              |
| 29 | 503056       | 514736        | 1747              |
| 30 | 522405       | 525915        | 1747              |
| 31 | 600392       | 669688        | 1921              |
| 32 | 609928       | 686072        | 1747              |
| 33 | 624184       | 691256        | 1921              |
| 34 | 635624       | 712216        | 1921              |
| 35 | 643336       | 652664        | 1747              |
| 36 | 667964       | 783556        | 1966              |
| 37 | 726104       | 796696        | 1921              |
| 38 | 802725       | 863835        | 1966              |
| 39 | 879712       | 901424        | 1966              |
| 40 | 898216       | 980984        | 1747              |
| 41 | 947835       | 1125765       | 1946              |
| 42 | 998104       | 1043096       | 1966              |
| 43 | 1077890      | 1099390       | 1966              |
| 44 | 1154450      | 1189150       | 1957              |
| 45 | 1156870      | 1292570       | 1946              |
| 46 | 1175265      | 1438983       | 1747              |
| 47 | 1185376      | 1286744       | 1929              |
| 48 | 1280565      | 1340235       | 1747              |
| 49 | 1328470      | 1483850       | 1966              |
| 50 | 1358595      | 1486845       | 1747              |
| 51 | 1392368      | 1464592       | 1747              |
| 52 | 1466150      | 1747930       | 1966              |

**Figure D.6** *(Continued)*

| | | | |
|---|---|---|---|
| 53 | 1468324 | 1749212 | 1967 |
| 54 | 1511930 | 1598470 | 1946 |
| 55 | 1669910 | 2062570 | 1966 |
| 56 | 1798875 | 1870245 | 1967 |
| 57 | 2082464 | 2090656 | 1747 |
| 58 | 2236570 | 2429030 | 1966 |
| 59 | 2652728 | 2941672 | 1921 |
| 60 | 2723792 | 2874064 | 1929 |
| 61 | 2728726 | 3077354 | 1966 |
| 62 | 2739704 | 2928136 | 1747 |
| 63 | 2802416 | 2947216 | 1747 |
| 64 | 2803580 | 3716164 | 1967 |
| 65 | 3276856 | 3721544 | 1747 |
| 66 | 3606850 | 3892670 | 1967 |
| 67 | 3786904 | 4300136 | 1747 |
| 68 | 3805264 | 4006736 | 1929 |
| 69 | 4238984 | 4314616 | 1967 |
| 70 | 4246130 | 4488910 | 1747 |
| 71 | 4259750 | 4445050 | 1966 |
| 72 | 4482765 | 5120595 | 1957 |
| 73 | 4532710 | 6135962 | 1957 |
| 74 | 4604776 | 5162744 | 1966 |
| 75 | 5123090 | 5504110 | 1966 |
| 76 | 5147032 | 5843048 | 1747 |
| 77 | 5232010 | 5799542 | 1967 |
| 78 | 5357625 | 5684679 | 1966 |
| 79 | 5385310 | 5812130 | 1967 |
| 80 | 5459176 | 5495264 | 1967 |
| 81 | 5726072 | 6369928 | 1921 |
| 82 | 5730615 | 6088905 | 1966 |
| 83 | 5864660 | 7489324 | 1967 |
| 84 | 6329416 | 6371384 | 1966 |
| 85 | 6377175 | 6680025 | 1966 |
| 86 | 6955216 | 7418864 | 1946 |
| 87 | 6993610 | 7158710 | 1957 |
| 88 | 7275532 | 7471508 | 1967 |
| 89 | 7288930 | 8221598 | 1966 |
| 90 | 7489112 | 7674088 | 1966 |
| 91 | 7577350 | 8493050 | 1966 |
| 92 | 7677248 | 7684672 | 1884 |
| 93 | 7800544 | 7916696 | 1929 |
| 94 | 7850512 | 8052488 | 1966 |
| 95 | 8262136 | 8369864 | 1966 |
| 96 | 8619765 | 9627915 | 1957 |
| 97 | 8666860 | 10638356 | 1966 |
| 98 | 8754130 | 10893230 | 1946 |
| 99 | 8826070 | 10043690 | 1967 |
| 100 | 9071685 | 9498555 | 1946 |
| 101 | 9199496 | 9592504 | 1929 |
| 102 | 9206925 | 10791795 | 1967 |
| 103 | 9339704 | 9892936 | 1966 |
| 104 | 9363584 | 9437056 | ca. 1600/1638 |
| 105 | 9478910 | 11049730 | 1967 |

**Figure D.6**

| 106 | 9491625 | 10950615 | 1967 |
| 107 | 9660950 | 10025290 | 1966 |
| 108 | 9773505 | 11791935 | 1967 |

**Figure D.6**

7. Pythagorean triples with a pair of palindromic numbers

| 3 | 4 | 5 |
|---|---|---|
| 6 | 8 | 10 |
| 363 | 484 | 605 |
| 464 | 777 | 905 |
| 3993 | 6776 | 7865 |
| 6776 | 23232 | 24200 |
| 313 | 48984 | 48985 |
| 8228 | 69696 | 70180 |
| 30603 | 40804 | 51005 |
| 34743 | 42824 | 55145 |
| 29192 | 60006 | 66730 |
| 25652 | 55755 | 61373 |
| 52625 | 80808 | 96433 |
| 36663 | 616616 | 617705 |
| 48984 | 886688 | 888040 |
| 575575 | 2152512 | 2228137 |
| 6336 | 2509052 | 2509060 |
| 2327232 | 4728274 | 5269970 |
| 3006003 | 4008004 | 5010005 |
| 3458543 | 4228224 | 5462545 |
| 80308 | 5578755 | 5579333 |
| 2532352 | 5853585 | 6377873 |
| 5679765 | 23711732 | 24382493 |
| 4454544 | 29055092 | 29394580 |
| 677776 | 237282732 | 237283700 |
| 300060003 | 400080004 | 500100005 |
| 304070403 | 402080204 | 504110405 |
| 276626672 | 458515854 | 535498930 |
| 341484143 | 420282024 | 541524145 |
| 345696543 | 422282224 | 545736545 |
| 359575953 | 401141104 | 538710545 |
| 277373772 | 694808496 | 748127700 |
| 635191536 | 2566776652 | 2644203220 |
| 6521771256 | 29986068992 | 30687095560 |
| 21757175712 | 48337273384 | 53008175720 |
| 27280108272 | 55873637855 | 62177710753 |
| 30000600003 | 40000800004 | 50001000005 |
| 30441814403 | 40220802204 | 50442214405 |
| 34104840143 | 42002820024 | 54105240145 |

**Figure D.7**